ENTROPIC SPACETIME THEORY

SERIES ON KNOTS AND EVERYTHING

Editor-in-charge: Louis H. Kauffman

K&E Series on Knots and Everything — Vol. 13

ENTROPIC SPACETIME THEORY

Jack Armel

The Foundation for Physical Sciences
USA

World Scientific

Singapore • New Jersey • London • Hong Kong

Published by

World Scientific Publishing Co. Pte. Ltd.

P O Box 128, Farrer Road, Singapore 912805

USA office: Suite 1B, 1060 Main Street, River Edge, NJ 07661

UK office: 57 Shelton Street, Covent Garden, London WC2H 9HE

Library of Congress Cataloging-in-Publication Data
Armel, Jack.
    Entropic spacetime theory / Jack Armel.
      128 p. 22 cm. -- (Series on knots and everything : 13)
    Includes index.
    ISBN 9810228422
      1. Space and time.   2. Entropy.   I. Title.   II. Series: K & E
series on knots and everything : vol. 13.
    QC173.59.S65A865   1996
    530.1--dc20   96-28661
                                   CIP

British Library Cataloguing-in-Publication Data
A catalogue record for this book is available from the British Library.

This book is printed on acid-free paper.

Printed in Singapore by Uto-Print

ENTROPIC SPACETIME THEORY —A COMMENT
by Louis H. Kauffman

The purpose of this short comment on Jack Armel's Entropic Space-time Theory (EST) is to bring out, in skeletal form, the major assumptions of the theory and how these assumptions work together.

First, it is assumed as given the de Broglie formula $E = h\omega$ where $\omega = c/\lambda$, h is Planck's constant, c is the speed of light, λ denotes wavelength and E is energy. It is assumed that the de Broglie formula applies to the energy of space itself.

Second, it is postulated that there is a quantum of space with minimal length l_0, and that to each (quantum of) energy there is an associated distance l. This distance l is an integral multiple of l_0.

Third, energy is stored in space according to Hooke's law: $E = kl$. (The constant k is determined by constraints of the theory).

In this context of space quanta, de Broglie and Hooke, it is natural to associate a wavelength λ with a length l: $l = \lambda x$. What is x? We have $E = h\omega = hc/\lambda = hcx/l$, and we have $E = kl$. Thus $kl = hcx/l$. Hence $x = kl^2/hc$ and we conclude that for λ the wavelength of a space-quantum of length l, $\lambda l = hc/k$. Wavelength is inversely proportional to physical length with a constant of proportionality hc/k.

Now we assume that the minimal length l_0 corresponds to the minimal energy $E_0 = h$. Thus $h = E_0$ implies $\omega_0 = 1$. This implies $\lambda_0 = c$. Hence $l_0\lambda_0 = hc/k$ becomes $l_0 c = hc/k$. Thus $l_0 = h/k$. The minimal length is the ratio of Planck's constant and the spring constant.

Now consider the energy density $\rho_0 = E_0/l_0{}^3$ associated with the space quantum l_0. We have $\rho_0 = E_0/(h/k)^3 = h/(h/k)^3 = k^3/h^2$. Planck's constant being very small, the energy density of the EST vacuum is very high! See the text to see just how high this density turns out to be. This high vacuum energy density is one of the key points of EST. Space itself becomes the final reference for all energetics.

So far we have made no assumptions about the geometrical nature of the space-quantum, other than its associated length l and elastic ability to store energy. EST further assumes a geometric model for these quanta as small chunks of spacetime. Model building for more complex quanta such as photons, electrons and quarks proceeds geometrically and topologically under the constraints imposed by known facts and parameters about these entities. Many remarkable fittings occur at the submicroscopic and cosmological levels.

PREFACE

Even if there is no end in the search of knowledge, I awakened this morning with a sense of completion. The last page was in place, the giant crossword puzzle was filled in...the entropic spacetime theory was as complete as I was capable to make it. Please don't laugh when I tell you that the morning after completing this work I started on two new aspects of this theory which I felt were not yet properly looked after. What kind of a theory is it?

It is a theory of space. It suggests how and when all matter and radiation was created. It proposes a cosmology. It delineates what "dark matter" is. It describes a discrete universe and offers specific dimensional limits.

It preserves the conservation of energy in all its aspects; there is no need to abandon it in cosmology. It has order—it cycles from expansion (where we are today) to an equilibrium, than to a compressive epoch. The universe never reached the state of "singularity" as many of our brothers have envisioned. The cosmos just "breathes" like we do.

The destiny of humanity is not forever, but we are likely to exist long enough to attain great ends. The projected length of our existence in this universe is about twice the age of the present epoch. That is, we should exist through the equilibrium and again as long as the universe is old now. The equivalent of some forty billion years of the projected future seems quite ample.

Yes, there was a big bang, the theory proclaims and there will be another and another in due course. The bang will be caused by matter—antimatter release of energy. Like the seasons, each new cycle will carry some of the previous seasons surviving galaxies into the new cycle. In other words each new epoch will start expanding fully equipped with the elements of the galaxies. An expanding epoch cannot create new galaxies from space. That is why our astronomers cannot place the creation of galaxies into our present epoch.

But then this theory does not deal with cosmology exclusively. It actually deals with the nature of space. I only told you about the future of our universe so that you will say—"It sounds all right but how did you figure this out?"

You will be surprised, as I was, when you find that space is very dense—more dense than any matter can be. You will be less surprised when I tell you that this theory says that time is variable and is not much good for measuring the passing of events for really long periods.

"How is it then that you gave predictions in billions of years?" You might say.

Well it happens that most of the expansion has already taken place (85%) in only half of the expected duration of the universe, before compression starts. Moreover the compression epoch starts very slowly and speeds up later. So if you believe that our present age is already some 20 billion years old (our time) then we will have another twenty billion before compression starts and a comparable twenty billion more in the next epoch—the one where antimatter is being created. In other words time crawls in tiny increments in the early stages after the big-bang. From now until the corresponding point in the compression epoch which resembles ours, time is not quite so ridiculous and can be approximated to units which we understand.

I have another surprise for you, before you ask how is matter or antimatter being created. In this theory, space, (called entropic spacetime) is the creator of all elementary things: radiation and matter. It can only do that in the compressive epoch.

There are other surprises in this theory. The universe is discrete and finite. Everything—not only energy comes in little bundles. Length, charge and time are all quantized.

Another surprise is that space is not a void or a vacuum but consists of tiny electric charges (a ten billionth the size of the electron). These charges are both positive and negative and are called dipoles. The distance between the charges is time and length. (Spacetime in effect). The energy content of space spreads the charges apart and hence we have expansion of space.

This makes it possible to explain gravity.

This is a highly deterministic theory—just the opposite of everything you've been accustomed to hearing about physics. There are numbers and values for all these items. There is no infinity, except in mathematics—and that is one of humankind's exclusive inventions which is not taken directly from observed nature. No, it does not attack or destroy the present physics of relativity or quantum theory. It sits in between the two, because it deals exclusively with space as a thing. It deals mostly with "rest energy" not kinetic energy. The latter is left to the present physics.

This covers the "WHAT". If you want to find out "HOW" then you'll have to read on.

Would you believe that it took more than thirty years to figure this out? The last ten rather intensively! I must be grateful to Helen for putting up with me for such a stretch. Also my son David, who calls me regularly to ask how the "great work" is coming along. Most of all I am grateful to Lou Kauffman, the professor and knot theorist from the University of Illinois, who could easily pass as my son. He has challenged and inspired me all at once and for a long time.

CONTENTS

CHAPTER I

AN OVERVIEW OF THE THEORY

If Niels Bohr were alive today and reviewed the Entropic Spacetime Theory (EST), he would probably say: "Not crazy enough!" And he'd probably be right.

EST is theory of space. It is based on a discrete universe—meaning everything is quantized, length, time, all things. There is no continuum!

Yes, there was a big bang, but it was caused by matter—antimatter annihilation. Sakharov first suggested this. He also suggested that gravitation is the metric elasticity of space. But he didn't put it all together.

John Archibald Wheeler, our living disciple of Bohr, interjects: "No point is more central than this, that empty space is not empty...Virtual pairs of positive and negative electrons, in effect are continuously being annihilated and likewise pairs of μ mesons, pairs of baryons and pairs of other particles. All these fluctuations coexist with quantum fluctuations in the geometry and topology of space".

This theory agrees that space is not empty but disagrees about the turbulence: Space is not a scene of locally violent conditions as Mr. Wheeler suggests. Rather, space is the ultimate "heat sink" of entropy. The energy in space is potential, unrecoverable in our cosmic epoch of expansion. This does mean that in the case of virtuals, EST may be a source of borrowing temporarily, but the energy must be returned. Kinetic particles in search for a "Higgs" particle create their own kind from the potential energy which IS space. Perhaps this is why the Higgs particle has not been observed.

Actually this theory is the child of George Gamow, another disciple of Bohr, who conceived the idea of the big bang (although he didn't name it so. It was Hoyle who gave it that name.) The inspiration from Gamow reads like this:

> "Just as no velocity can exceed that of light (c), no mechanical action can be smaller than the elementary action (h) no distance can be smaller than

the elementary length (ℓ) and no time interval can be shorter than the elementary duration (ℓ/c). When we know how to introduce ℓ and ℓ/c into the basic equations of theoretical physics, we will be able to state proudly: Now at last we will understand how matter and energy work".

Gamow's projection for the future of physics is a simple one. Unfortunately he doesn't say how this is to be accomplished. Albert Einstein, the great genius of the twentieth century, placed the two theories (special and general) of relativity on the table pretty independently. Yes, he leaned a little on two mathematicians to help him from time to time (Minkowski and Friedmann), but essentially it was his baby. The other great theory, Quantum, came in spurts and starts by way of deBroglie, then Schrödinger and Heisenberg, again fathered by Bohr, and a whole German team of lesser genii—but genii nonetheless. The world of physics became totally enveloped in probability, waves and mathematics.

Louis Crane, the mathematician, wrote to me recently: "Modern physics has led us into a region far from our experience where common sense in no guide and ordinary language loses its meaning. We navigate through this region only through use of the compass of abstract mathematics." This has been said in similar tones by both Schrödinger and Heisenberg.

There is nothing wrong with probabilities. Einstein, however spent his later years in pursuit of a theory which would offer more determinism. Looking back at this schism, P.A.M. Dirac had this to say about it:

"Some physicists led by Einstein have supposed that physics should be deterministic and should not merely provide probabilities. Now Bohr accepted the probability interpretation...and that led to very big controversy which lasted throughout Einstein's life. The question is who is right of the two?"

"It seems that according to the standard accepted ideas of atomic theory, Bohr was right. The interpretation in terms of probabilities, based on the Schrödinger wave function, is the best we can do...According to present quantum mechanics, the probability interpretation which was championed by Bohr is the correct one. But I think that ultimately Einstein will prove to be right, because the present form of quantum mechanics should not be considered the final form...It is the best that one can do until now. But one should not suppose that it will survive indefinitely into the future. And I think it is quite likely that at some future time we may get an improved quantum mechanics in which there will be a return to determinism and which will therefore satisfy the Einstein point of view. But such a return to determinism could only be made at the expense of giving up some other basic idea which we now assume without question."

When one gives thought as to what the two great theories have in common one finds little.

Richard Feynman says: "The quantum mechanical aspects of nature have not yet been carried over to gravitation...For consistency in our physical theories it would be important to see whether Newton's laws, modified to Einstein's law can be further modified to be consistent with the uncertainty principle. This last modification has not yet been completed."

In other words the two theories do not have mutual consistency. There is a gap which needs to be covered. If they have little in common, one could ask: What is it that these two theories seem to have missing in common?

AN APPROACH TO SPACE

Einstein proclaimed that space is a continuum of void and therefore need not be considered in the workings of bodies of matter. In other words space is not a thing! Yet we have fields in space.

Richard Feynman said once in one of his lectures: "I have no picture of this electromagnetic field...When I start describing a magnetic field through space and I speak of E and B fields and wave my arms you may imagine that I see them". Feynman admitted quite candidly that he didn't have the slightest idea of what a field really is.

What do we have on the positive side of a Space Theory?

The famous Michelson-Morley experiment which demolished the ether concept of space, actually was the only experiment which gave space a property which was unknown before. It proved that space will allow light to travel from a moving body at an equal velocity in any direction regardless of the velocity of the body.

Now we have a property of space! We have Dirac's idea of space which is full of charges. We have Wheeler's idea of an unempty space full of turbulent virtuals of every kind, fluctuating at $\sim 10^{94}$ g/cm^3. To this he added:"...For this multiple-connectedness of space of submicroscopic distances no single feature of nature speaks more powerfully than electric charge."

Then Feynman has something to say: "Perhaps gravitation and electricity are much more closely related than we think." To this he adds "No machinery has ever been invented that 'explains' gravity without also predicting some phenomenon that does not exist". Does he mean exist ever? Or does he mean, yet exist?

So we have some output leading to a space theory but we still don't have a theory.

A CURE FOR INFINITY

When space was considered a vacuum there came certain problems to our quantum theorists. Let us consider the problem of infinity and space. A Quantum Field Theory was devised, which in part is explained this way.

A field contains an infinite number of oscillators so that energy can mount up to an infinite amount. But this is something they, the theorists can live with. They measure energy as a difference from one state to another. So there is no energy in the vacuum. A field can have an infinite amount but space contains zero.

Another infinity problem arose in QED (quantum electrodynamics). If an electron were ideal and went from point to point in space by a direct path there would be no problem. Using QED terminology, n would be the mass of the electron and j would be the charge. Unfortunately no such electron exists. The observed electron emits and absorbs its own photons from time to time. Since the mass and charge of an electron are affected by various alternatives expounded in the QED, the experimentally measured mass m and the experimentally measured charge e of the electron are different from n and j. If there were a definite mathematical connection between n and j on the one hand and m and e on the other there would be no problem. But there isn't.

This is how Richard Feynman describes the calculation for m.

"We write a series of terms...The first term has no couplings and represents the ideal electron. The second term has two couplings and represents a photon being emitted and absorbed. Then came terms for four, six and eight couplings, etc. When calculating terms for couplings we must consider all possible points where couplings can occur, right down to cases where two coupling terms are on top of each other, with zero distance between them. The problem is when we try to calculate all the way down to zero distance the equation blows up in our faces and gives meaningless answers—like infinity."

"Well instead of including all possible coupling points down to zero, if one stops the calculating when the distance between coupling is very small, say 10^{-30} cm, then there are definite values for n and j we can use so that the calculated mass comes out to match m observed in the experiment and the calculated charge matches the observed charge e ."

This is all very well since a tiny space was arbitrarily selected—a tiny space, perhaps of the kind Gamow had in mind for ℓ.

"If someone else comes along stops calculations at a different distance, say 10^{-40} cm then their values for n and j which are needed to get m and e come out differently.

How was this resolved?

It was observed that if two theorists stopped at different distances to determine n and j and then calculated the answer to some other problem each using appropriate (but different) values for n and j the answers came out nearly the same. Feynman, Schwinger and Tomonaga finally invented ways to make the calculation confirm. (The n and j values are purely theoretical and are not directly observed anywhere). Feynman himself found this process (called renormalization) mathematically unsound. Feynman called it "dippy".

Renormalization is standard procedure today! If this effort were used to find the Gamow number for ℓ, then, once determined would have eliminated infinity from QED.

It might be of interest to state that all efforts to renormalize values for the gravity equations do not work. That is why Feynman said that the modification of the Einstein law with quantum theory has not been completed.

At this point there is a great temptation to spring a value for the smallest length ℓ, as developed in this theory. With great restraint I am refraining from this so it can be revealed at the proper time and hopefully leading the reader into the pages which follow. What has been revealed so far is that a search for a firm number is possible in quantum theory. This is perhaps the first step in the determinism which Einstein sought. All is not based on probability.

CONSERVATION OF ENERGY

The theory of entropic spacetime deals with one of the most fundamental laws of nature, the conservation of energy. When we get into cosmology we find that some stray from time to time. The cosmologist E.R. Harrison leaves no doubt about his position on this point.

> Science clings tenaciously to concepts of conservation, the most fundamental of which is the conservation of energy principle. Whenever scientists have found that energy has vanished...they have searched diligently for it reappearance in some other form. The conservation of energy serves us well in all sciences except cosmology...but in the universe as a whole it is not conserved. The total energy decreases in an expanding universe and increases in a collapsing universe. To the question where the energy goes in an expanding universe and where it goes from a collapsing universe the answer is—nowhere, because in this one case energy is not conserved."

When I asked Dr. Morton Silberstein what he thought of this argument, his reply was: "How does it (the conservation law) know when to cease operation?" When I asked Prof. M.J.G. Veltman of the University of Michigan that same question he replied that it goes into the curvature of space. These were pretty snappy answers!

EST provides a clear place for the energy to go. Moreover the "lost" energy, or the entropy, becomes the cause for the expansion of space. In a contracting epoch of the universe the released energy becomes the building block of everything; Radiation and matter.

QUANTIZATION OF SPACE

Dirac in his work of combining relativity with quantum theory suggested that space is not zero but a negative energy made up of positive electrons. He saw the value of having his charges making up the very nature of space; if he didn't quantize them, it seemed unnecessary—the positrons were already quanta of energy. Dirac maintained in his lectures that the electron charge, being everywhere equal, whether positive or negative should be the basis of quantization.

It would only be proper to remind the reader that Dirac is the one who first anticipated the matter and antimatter principle. In the time he speaks of positrons it was some time after their discovery. This discovery was due to his prediction. He admits shyly, that when he first thought of space as filled with positive particles he actually thought of protons, but he was too timid to say this. When the positron was discovered he found his positive particle for the space concept.

His logic continues:

> "The more electrons we put into states of negative energy, the lower the total energy becomes...Thus we must set up a new picture in which all negative energy states are occupied and all positive states are unoccupied."

Then he focused on the "holes" in the vacuum; this "hole" will be a positive energy. From this he concluded:

> "Previously people thought of the vacuum as a region of space completely empty, a region that doesn't contain anything at all. We may say that a vacuum is a region of space where we have the lowest possible energy."

In general, this concept of Dirac was rejected. Probably the prime reason was that negative energy was not only never observed and the concept threatened to turn all of physics upside down.

Actually the idea was brilliant—it only needed a few adjustments. Quanta of charge was brilliant but the electrons were much too big. They needed to be 10^{-10} smaller than electrons and they had to be positive energy, not positive charge. The sentence which speaks of space as the region of the lowest possible energy was, in our view, right on!

These tiny charges had to be energy too—just as electrons are energy. Space now took on a new look. We have quanta of energy charge being space.

There was one more requirement. The charges had to be neutral rather than negative or positive. So we made them dipoles. Yes there is still another requirement!

Space is expanding in our present cosmology. This means that the energy charge (like spacetime) was much smaller once and is now in a state of expansion. If each quantum of space would expand by receiving energy from the decaying kinetic system—the potential energy which we call entropy—would accomplish two things. One would be to account for the expanding cosmology, the other would be to introduce another dimension to the space quantum, namely length ℓ. This length would be proportional to the energy charge.

Strangely we find Gamow's length a variable, depending on the age of the universe.

Now we can't help the QED boys in finding the length ℓ_o for their renormalization. (But we may yet)

We can however establish a time interval ℓ/c. We will find that time is a variable too; this is in agreement with relativity and extends the value of the time interval to the age of the world. Soon we'll be floating in deterministic values so rare in quantum theory.

DILATION

A quantum that varies with the age of the universe should have a ℓ_o at the time of the big bang, a ℓ_{now} at present and ℓ_q at the time equilibrium, if there is such a thing. The latter depends on the energy content of space itself. If this contribution is high enough we should have a cycling universe.

So far we have the main attributes of EST:

- Neutral electrically with a new kind of tiny charge
- Indetectable because of its size and ubiquitous character
- Consists of charge, energy, distance and time

From the viewpoint of quantum theory, EST is different. We are not dealing with location and momentum which obeys the Heisenberg doctrine. We are dealing with something scalar and ubiquitous. In a universe of bosons and fermions we have to select a new type of boson as the character of a space quantum. We'll call them scalar bosons and say that their spin is zero, compared to radiation as vector bosons spin 1.

CHAPTER II

THE FABRIC OF
ENTROPIC SPACETIME (EST)

According to the EST theory, the spacetime of relativity is incomplete without adding the attributes of energy and charge to it and quantizing the whole business. Each quantum of EST has a finite quantity of E, energy of q, charge, of ℓ, length and a time interval related to length ℓ/c.

These four attributes of a EST quantum are interdependent so that an increase in say, energy automatically increases the charge, and length and the time interval. An increase in energy content is due to the entropy of the universe and hence we have the expansion of space.

An increase in energy of an EST quantum is the equivalent of stretching the quantum—thus storing potential energy in it. In an expanding universe we have the cause for the unrecoverable energy of entropy from all mechanical actions— the second law of thermodynamics. We also have the cause for gravity—a space fabric in tension.

We have basis for believing in the conservation of energy. The total energy in the universe is fixed. Its distribution varies.

The EST quantum is described as a scalar boson which operates in two dimensions and time—a total of 3 dimensions, as compared to vector bosons and mass particles (fermions) which operate in three dimensions and time; a total of four dimensions. In a universe which is in compression, (rather than the present expansion) EST quanta lose energy and become the building blocks of photons and matter through topological transformations (which are described in Chapter VI). In the language of quantum mechanics the scalar boson has a spin of 0, as compared to vector bosons with a spin of 1 and fermions with a spin of 1/2.

Why do we think the universe will cycle? When we get to evaluate the quantity of energy which is in space, there should be no doubt about the needed gravitational capacity to cycle.

The absence of vectors gives the scalar bosons a special character. Unlike particles in quantum mechanics, the scalar bosons don't wiggle and hence Schrödinger's ψ is not applicable. Moreover as we shall see when values for E are deduced, the energy content per boson is of such low value that their temperature (E/k) is cryogenic and well below the lambda point. ψ becomes applicable when there is a momentary interaction between scalar boson and particles as in virtual particles. Otherwise the creation of photons (E/c) and fermions (E/c²) can only be generated when the universe is in the compressive epoch—while we are today in the expansive epoch. What we see today in apparent generation of matter is from the degeneration of larger existing particles in a downward cascade. According to this theory matter cannot be created from the scalar bosons of EST in this epoch. The scalar quanta are defined as $E/c^2 . 1/\mu_o \epsilon_o = E$.

Without momentum or "mass" the determination of dimensional values for scalar bosons could be defined dimensionally without conflict with Heisenberg. We are never in the condition of searching for location while needing the momentum of the scalar boson. Their location is ubiquitous while they have no momentum.

A QUANTITATIVE DESCRIPTION OF SPACE

We will seek values for energy E, charge q and length ℓ at their minimum value and at their present dilated condition. The time interval t varies as ℓ/c. E, q, and ℓ are covariant and directly proportional.

Sources of basic dimensions:

- The Planck equation for length is assumed as the basis for EST quanta today— that is in its dilated condition. The reason for this is that it contains G, the Newtonian "constant" which has been subject to much measurement. G has never been found directly connected with the prime constants of h and c except through the Planck length ℓ. G could therefore be considered a variable and a measure of "stretch" of space.

- The specific charge of the electron, $-e/m_e = -ec^2/E_e$ becomes the basis for the total charge in the EST scalar boson. The total charge is neutral in the scalar boson and the opposite charges are separated by the time interval. The discernible charge such as that of the electron becomes measurable only after topological transformation.

- The numbers obtained above must be in agreement with the classical electro-static equation $E = 1/2 \cdot e^2/r$ so that

$$\ell_o = \frac{q_o^2}{2E_o}$$

The minimum energy cannot be smaller than h, the Planck constant.

■ The scalar bosons do not require $\hbar = h/2\pi$. As we shall see in the topological model making of photons, electrons, protons and neutrons the 2π is one of the vectors used in model making.

$$\ell_P = \left[\frac{\hbar\,G}{c^3}\right]^{1/2} = 1.61605(10)\times10^{-35}\text{m}$$

$$\ell_{\text{now}} = \left[\frac{h\,G}{c^3}\right]^{1/2} = 4.0508359\times10^{-35}\text{ m}$$

In addition to the above requirements, scalar bosons have to fit with wavelengths and frequencies which are the dimensions of photons and nucleons which can be created from the scalar bosons, when the universe is in the compressive state. The frequency, when in the scalar state represents dilation, or stretch or tension. This may be visualized as a string in tension—when the vector of "plucking" is added then λ and ω become evident. In our expanding universe this tension is merely potential energy. The frequency is a multiple of E_o.

In accordance with this theory, space is three dimensional—one of these dimensions is time. One might surmise that there are two other dimension. Actually, we give only one other dimension, the length ℓ and call it gravitational length.

TIME

SPACE

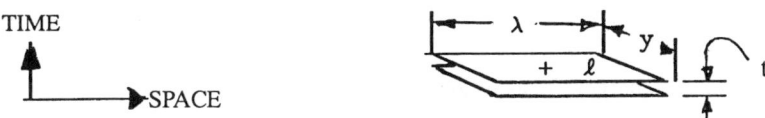

FIG. 1. A quantum of space

In Fig.1 we see a concept of two planes of opposite charges separated by time. The dimension ℓ appears to be more an area than a length. Here ℓ is defined as a product ($\ell = \lambda y$). λ is in meters by y is a unitless dimension. Thus ℓ is given in meters. To distinguish it from an ordinary length we call it a gravitational length, for lack of a better name. It is possible that in a universe which we consider discrete, a line has to have a thickness lest we accept the euclidean idea of a line which only has one dimension and therefore its thickness is infinitely thin. There is no infinity in this concept.

In the model making stage we will find that λ becomes the wavelength and y becomes the amplitude of a vector boson wave. As scalar bosons of EST the best we can do is to call them potential wavelengths and amplitudes. In space we only have the stretch. To become waves we must have twist as well. In EST all components are there with the exception of twist (angular momentum). This relates further to the fact that we are in an expansive epoch of the universe—hence space is all tension and cannot twist from its own potential energy.

External energy, from the kinetic system can be the cause of virtual particles, and also cause curvature and gravitational waves.

Below are listed specifications giving values:

The rest energy for the minimum scalar boson E_o, the present scalar boson E_{now} and for comparative reasons, the rest energy for an electron and a proton is given below:

$E_o = h = 6.6260755 \times 10^{-34}$ J

$E_{now} = 2.115509 \times 10^{-23}$ J

$E_e = 8.1871113 \times 10^{-14}$ J

$E_p = 1.5032787 \times 10^{-10}$ J

The values of the gravitational lengths ℓ are as follows:

$\ell_o = 1.2687795 \times 10^{-45}$ m

$\ell_{now} = 4.0508359 \times 10^{-35}$ m

$\ell_e = 1.5676911 \times 10^{-25}$ m

$\ell_p = 2.8785203 \times 10^{-22}$ m

The meaning of gravitational length, as distinguished from length in meters has a basis in the two dimensional character of EST. As energy is added to a EST quantum the quantum changes in two dimensions λ & y. λ becomes smaller and y enlarges. $\lambda = hc/E$, $y = \ell E/hc$.

The values of charges are as follows:

$q_o = 1.2966903 \times 10^{-39}$ C

$q_{now} = 4.1399468 \times 10^{-29}$ C

$-e = 1.6021773 \times 10^{-19}$ C (also known as the elementary charge of the electron)

$q_p = 2.9418424 \times 10^{-16}$ C (represents the total scalar charges, which are mostly neutralized by the topological arrangement, exposing only $+e$)

The potential frequencies of minimum and present scalar bosons are as follows:

$\omega_o = 1$

$\omega_{now} = 3.1927028 \times 10^{10}$

The actual frequencies of the electron and proton:

$\omega_e = 1.2355898 \times 10^{20}$ Hz

$\omega_p = 2.2687316 \times 10^{23}$ Hz

$\omega_{now}^2 = y_{now}/y_o$ while $\omega_{now} = \lambda_o/\lambda_{now}$

THE CONSTANT OF ELASTICITY:

Using Hooke's Law which states that stress is proportional to strain permits us to create a constant of dilation, $k_{dil} = E/\ell = 5.22240(11) \times 10^{11} Jm^{-1}$.

DESCRIPTION OF SPACE:

Today's space quanta are a ten billionth in size of the electron and the electron is the smallest particle we know. Today's space quanta are 10^{10} larger than the space quanta were at the time of the big bang. This incredible smallness and accordingly low temperature does not prevent the EST from being the densest of all things.

It is conceivable that in a 3D map of the universe we could have a considerable variation in the values of the scalar bosons. Scalar bosons in the proximity of large bodies (galaxies) which are decaying into radiation ultimately end up as entropic energy which enlarge the nearest space quanta. It can be likened to pulling on a fishing net—the nearest "square" becomes distorted, stretched, and the stretching is passed on to its neighbors. This distribution of energy is passed on at velocity c, just as photons are. This is considered the reason for the expansion of the universe.

Even the high velocity of the distribution of energy is dwarfed by the vast volume of the universe and hence there might be some lack of uniformity in the energy content in different parts of the universe. This corresponds to the curvature of space from relativity. As Einstein said: "masses tell space how to curve space and space tells the masses how to travel." Curvature of space is the geometric equivalent of the topological distribution of energy in space. What this theory adds is the abolishment of the continuum aspect of space, it adds quantization of space, and provides the basis for a finite universe.

Remote space can be defined as the most dense and it would seem to be so at about 3 1/2 kiloparsecs away from the nucleus of the nearest galaxy. The need for dark matter was deduced from data by Vera Rubin and the role of EST as dark matter is described in greater detail in *A Rationale for Dark Matter*.

As a fabric of space we must describe how this "fishnet" is multiply connected from scalar boson to scalar boson and to matter. This net could be visualized as made up of stretched rubber bands. Where mass particles occur we visualize knotted twisted up balls. The entropic energy flowing from the mass to the network continues to be distributed to the remainder of the EST at the velocity of light. Motion of masses can be visualized in this network without ever breaking the bond between the mass and the fabric. Since all masses move at velocities less than c and the distribution of the energy is at c there is no need to fret about this bond breaking.

What about a mass spinning? Dirac has already shown that such a bonded and spinning mass will restore the network at every 4π revolutions to its normal relationship. There is no possibility that such spinning will "tear the coordinates" apart.

A photon, which is a spin 1 boson, can be described as a torsion wave in the fabric itself. Energy is transported through the medium of the fabric without anything actually "moving".

When a mass is in motion in the fabric and emits a beam of light, we should find the velocity of the light equal both in the direction of the path of the mass and in the opposite direction as well, (light as a torsion wave propagates from scalar boson to scalar boson at velocity c).

The abandonment of space (as ether) as the medium for the travel of light which resulted from the Michelson-Morley experiment can actually be used as support of the EST theory.

The term *fabric* used in this description should not be taken literally as to the actual existence of interwoven threads. It is better thought of as granular and minute groupings of energy-charge which are by this definition space. These groupings don't take up more or less space—they ARE space. How much energy charge comprises a granule of a scalar boson can be described in length ℓ, or in energy E or in charge q. They are all attributes of space. There is no space which does not contain EST—according to this thinking. Its dimensions may vary depending on the degree of expansion or compression of the universe. The conserved fixed amount of energy in the universe is divided between the EST and the rest energy of particles plus the energy in the kinetic state. The latter is the momentum of radiation and the momentum of mass particles in motion.

In the expansion epoch such as we are in now, energy flows into the EST. During compression epoch EST surrenders energy by the creation of particles.

To find out what a time interval is one may divide ℓ/c. Obviously time can also be defined in terms of E and q since they are convertible into one another.

Using this theory we can describe space as elastic, quantized and subject to stretch. We can state that the universe is cycling and finite. We can say that all particles exist because they have been subjected to twist (angular momentum). We have identified "dark matter" as being space itself and eliminated the idea of a continuum. Model making of photons, electrons, neutrinos, neutron/protons is possible.

From the description above it becomes plain that the overall dimension of a space quantum is the constant of elasticity multiplied by time (t k_{dil}). Time becomes the only variable which affects the overall dimension. Of course ℓ increases as E increases but their increase is proportional and therefore constant.

Another way we could describe this relationship is to say that with the expansion of space, ℓ increases in a spacelike fashion and time increases in the timelike fashion. The terms spacelike and timelike are not alien to relativity. In the light-cone of relativity, an object moving at a high velocity crosses fewer time lines.

The age of the universe cannot be properly measured by time. Rather than time, the age can be better defined by the number of times a scalar quantum has expanded from the minimum to the present. The theory suggests an equilibrium, when the universe will stop expanding, thus reversing the second law of thermodynamics. The relationship of time and space would not reverse but change direction by $\pi/2$ when the universe begins to contract. All of this and more will be detailed in subsequent pages.

CHAPTER III

GRAVITY

We can safely say that gravity applies to all things which contain energy. We may credit Eötvös with saying that gravity forces (or potentials) are proportional to energy content. Since all bosons and fermions contain energy then gravity applies there. The moot question is whether there is anything which doesn't contain energy?

If it is assumed that there is a ubiquitous field throughout the universe and that the vacuum is not empty there might be the assumption that it is filled with something. In this theory we go further. Space actually *IS* this universal field of scalar bosons and there is no other "empty" space. One need not ask "what is beyond" this field. This field is unique in character—it encompasses the entire universe.

Richard Feynman in his pursuit of a gravity theory (*Lectures on Gravitation 1962-63*) considers the spin 0 theory on gravitation and rejects it:

"We shall give an argument by analogy...We ask the question what is the attraction between moving objects; Is it larger or smaller than for static objects? We may for example calculate the mutual attraction of two gases; The experimental evidence on gravity suggests that the force is greater if the gases are hotter...In electrodynamics the electric forces are unchanged by random motions of the particles. Now the interaction energy is proportional to the expectation value of the operator γ_t which is $1/(1-v^2/c^2)^{1/2}$. Since the potential resulting from this operator is not velocity dependent, the proportional factor must go as $(1-v^2/c^2)^{1/2}$.

This means that the interaction energy resulting from operator 1, corresponding to spin 0 field would be proportional to $(1-v^2/c^2)^{1/2}$.

In other words the spin 0 theory would predict the attraction between masses of hot gases would be *smaller* than for cool gas...Thus the spin 0 theory is out—we need spin 2 in order to have a theory in which the attraction will be proportional to the energy content".

In the case of EST, the analogy is not the same as Feynmans hot gases. The relationship of EST and the kinetic items are multiply connected. A movement of a particle, which is "attached" to the space, stretches the space as if a hand was pulling on an elastic net. The dimensional distortions on the net are distributed across the entire network. The energy is distributed at velocity c just like light. The moving "hand" is driven by kinetic energy. The elastic net is yielding and reshuffling its potential energy. There is an "opposite" character to this relationship with that of the hot gases.

It is not Dirac's negative energy either—it is positive energy which maintains the field and determines the energy content of the scalar boson which is space itself. Moreover, EST is not considered a moving object (kinetic as in Feynman's example). This allows the spin 0 to be applicable to the entropic spacetime theory. The operator previously mentioned by Feynman becomes inverted when viewed from the EST outlook.

The entropic spacetime INCREASES in energycharge when it is expanded but becomes less dense in the process. The spin 0 theory rejected by Feynman as applicable to kinetic space (3 dimensional + time) is precisely applicable to the entropic spacetime theory. (2 dimensional + time)

This theory proposes that everything in the universe contains energy. Scalar bosons correspond to Spin 0, Vector bosons correspond to Spin 1 and fermions correspond to spin 1/2. Vector bosons and fermions belong to the "kinetic system". Scalar bosons belong to the entropic spacetime system.

The key to the visualization of entropic spacetime is that it stretches—it is elastic. We can pack more and more energy—charge into it and its volume density reduces.

GRAVITATIONAL ATTRACTION—WHY NOT REPULSION?

In the present concept of physics, the idea prevails that an increase in energy content also indicates an increase in density. The proton was always considered more dense than the electron for that reason. Yet by our method of reckoning, it turns out to be just the opposite. Feynman's example of the greater gravitational attraction of hot gases compared to the same gases at lower temperature is, of course, perfectly correct. The hot gases contain more energy and therefore they are more susceptible to a greater gravitational force.

Here is where the entropic space theory provides a simple answer.

We must first of all think of the entropic space as not only more dense than anything we call matter, but also that all dimensions of EST are smaller than those of matter particles. From this we can explain why gravity is always a single acting force and there is no antigravity.

The gravitational attraction between any two bodies:

$$F_G = \frac{E_1 E_2 \, \ell^2_{now}}{hc \, \ell^2} \quad (eq.1)$$

Where E is the rest energy of bodies 1 and 2. ℓ_{now} is the gravitational length of the space quanta and ℓ is the distance between bodies (usually called R).

ℓ_{now} is a tiny value - (4.05×10^{-35}) and would always be smaller than any distance between bodies. This is why gravity is always a single directional force.

The derivation of eq.1 is given below.

From $F_G = \dfrac{G \, m_1 m_2}{R^2}$ substitute for $G = \dfrac{\ell^2_{now} \, c^3}{h}$ from Planck length eq.

Substitute E/c^2 for m:

$$F_G = \frac{\ell^2_{now} \, c^3 E_1 E_2}{\ell^2 \, hc^4} = \frac{E_1 E_2 \ell^2_{now}}{hc \, \ell^2} \quad (eq.1)$$

From the above, we can see that the gravitational force between any two bodies is caused by the difference in the space dimension ℓ_{now} and the distance between the bodies. They are also proportional to the rest energy E of the bodies involved.

The ℓ term can be exchanged, to any term of EST dimensions such as charge or time, but this example is sufficient. The principle is that minuteness of space quanta makes them also more dense than any matter.

In the case where the gravitational force between two space quanta were sought, the distance ℓ would be the same as the ℓ_{now} and they would cancel out.

The Newtonian "constant" G has been eliminated in favor of the Planck length, which was adjusted to be $\ell_p.(2\pi)^{1/2} = \ell_{now}$.

The theory states that ℓ is a variable and gets larger as the universe expands. This makes G not a constant.

All particles want to be as small as possible; all energies want to be as small as possible. The two bodies, with the rest energy of E_1 and E_2 are bulging in the fine grid of E_{now} of entropic spacetime, and forces are acting from the entropic space against the two masses.

If this theory is correct, then we know that gravity is not a property of mass but rather an attribute of the relation of entropic spacetime interacting with the corresponding dimensions of bodies. Considering that EST is the parent of all matter then GRAVITY IS THE PROPERTY OF EST.

THE INTERFACE BETWEEN ENTROPIC SPACETIME AND THE KINETIC SYSTEM

The entropic spacetime theory opens the way to bring gravity into the quantum theory. Each space quantum has a group of neutral charges located on the surface of the quantum. These charges want to contract and are prevented from so doing by the energy content of the quantum. The electric force F_E is the element which attracts them. Its value is 1.517038×10^{-13} Newtons at the current state of the space quanta. The opposing force is calculated the same way and therefore is apparently equal to F_E. If we place a minus sign in front of the contracting force $-F_E$ then we have what would appear to be a perfect equilibrium and therefore a static (nonexpanding) universe. But this is not quite so. We know it is expanding! How are we going to explain this?

The answer is quite simple. We know that the electric force of attraction is very similar to the gravitational force except it is much larger. When we compare these forces using classical means we find the electric force is 4.16×10^{42} times greater. Now this is an extremely small number when compared to the value of F_E. If we were to try to add this number to F_E we would need at least 42 places after the decimal point to show that there is a difference. As you can see we have the value of F_E only to the seventh place. In effect, we have two almost equal forces working against each other with the difference being so tiny that the gravitational force is quite negligible between any pair of space quanta; for that matter it is negligible in the entire field of subparticles. It is just barely discernible using laboratory equipment and that is why it is so difficult to get a good number for G. But when we get to planetary masses then it becomes an outstanding force with which we are all familiar. Nonetheless, the gravitational force is part and parcel of the entropic electrical force F_E. It is the same force and has the same characteristics, except that it is also opposed by an almost equal force of the same nature.

Since the theory states that masses are also made up of charges, energy and dilation just as the entropic space, then we also have a gravitational force working in each proton, neutron and electron (but negligibly), each molecule, each cluster of molecules all the way up to each macro-mass. They all want to be as small as possible. When the masses are big enough, such as our planet, we have gravitational forces trying to contract it all the way to the center of the mass of our planet.

THE ACCELERATION OF GRAVITY FROM ENTROPIC DATA

On earth there is a familiar number called the acceleration of gravity often denoted by g.

Let's look at the G equation in terms of entropic data:

$$G = \frac{\ell^2_{now} c^3}{h}$$

If we now take the Newtonian equation for calculating the rate of acceleration g which is

$$g = \frac{G M_E}{R^2} = 9.8 \text{ ms}^{-2} \qquad = \frac{\ell^2_{now} c^3 M_E}{h R^2}$$

The M_E is the mass of the earth (6×10^{24} kg) and R, the radius (6.4×10^6 m)

The Newtonian G is gone! It becomes obvious that as bodies acted upon by gravity, the force is continuous and therefore the bodies will accelerate.

THE MACH PRINCIPLE

The Mach principle is defined as follows:

"All inertial forces are determined by and are proportional to the total amount of matter in the universe."

Up until now we have dealt only with the potential aspect of EST, so that all reference to masses or energy were references to rest masses. Once we get into dynamics we have to define the kinetic energy due to the motions involved. This brings us into the area of relativity and the need to merge the concepts so they are mutually in harmony.

Let's see what that really means; the concept has a long history which we will but touch and then only in the sense where it has bearing on our thesis.

From Newton we have:

"Absolute space in its own nature and without relation to anything external remains similar and immovable."

The concept of absolute space was demolished by relativity. But in the earlier thinking of Einstein the Mach principle influenced him. He too stated that local inertial forces are determined by the distribution and quantity of matter in the universe. This concept connects local inertial forces to the overall properties of the universe. Einstein summed it up:

"In a consistent theory of relativity there can be no inertia relative to space, but only inertia of masses relative to one another."

Soon after that, the mathematician Minkowski, at Einstein's side, pointed out that the equations already encompass the Mach principle independently and therefore the nature and existence of spacetime is not dependent on the existence of matter.

To further show the wavering positions taken by our greatest scientists on this subject I am going to quote Einstein where even he speaks from "both sides of his mouth." In the same book (*The Meaning of Relativity*, 5th Ed. Princeton) he states: "Mach's idea gains probability as it is an unsatisfactory assumption that inertia depends in part on mutual action of masses and in part on the independent action of space." Yet in the famous proof of the bending of light Einstein states: "half of this deflection is produced by the Newtonian field of attraction, the other half by curvature in space."

Still impressed by Mach, Einstein also stated:

> "If I have sufficient distance from all other masses in the universe, its inertia must fall to zero."

If the motion of each individual body is determined by all others in the universe then we have to ask the question of how this is accomplished through void space? Again we meet the old story about the "field" which enables properties of matter to be extended remotely through a vacuum.

All in all general relativity does not provide a formulation as to what the inertia of a body is, but at least we have in relativity the same treatment given to gravity as to inertia.

How would the entropic spacetime theory deal with questions of inertia, gravity and Mach's principle?

Let's imagine the effect of EST with masses in motion. The EST is static—do we have a flow of EST past a moving body?

Decidedly NO! It was already pointed out that if we grabbed a handful of EST fabric we could move our hand in any direction. In the process we would momentarily dilate the fabric behind the direction of the motion and relieve the tension in the direction of the motion. We know that dilation requires energy and therefore the converse, compression yields energy. This means there would be a flow of energy from the front to the rear of the motion. Since any energy in the EST moves at the velocity of light, and this movement takes place from quantum to quantum regardless of the actual dilation of each quantum. The time to move from quantum to quantum would be ℓ_{now}/c. Compressed or dilated, the velocity is c. Even though energy flows from front to back, the quanta don't move any more than the rubber band molecules flow past the object which does the stretching. This means that the relative position of the quanta are unchanged—only the energy content and the dilation has changed momentarily.

There is clearly a limit to any body's velocity in the EST. When the quanta are compressed to the lower limit of their dilation (ℓ_o) the charges are as close to each other as they can be and there is no way they can be compressed any further. Moreover this very high velocity body is so near the velocity of the energy transfer from front to back that the energy becomes part of the moving body and so the mass of the body increases as the relativity theory demands. The forces acting on the body also shorten the body's dimensions, again in agreement with relativity. The entropic spacetime theory allows us to visualize the phenomena of relativity. Even the slowing of time is accounted for since the dilations are decreased through compression and time is a function of dilation.

As to inertia of matter: Inertia was always described as a property of mass. In the entropic spacetime theory we see that it is caused by the space-mass relation. In the motion of a body, the compression of the EST only takes place where a nucleus is in direct confrontation with a space quantum "network". The distance between the quanta is very close and the nuclei are relatively very far apart. The reason why a greater force is required to move a more massive object becomes almost self explanatory. The greater the mass of the matter the greater the total number of compression-dilations occur. Hence the greater the mass the larger the inertia.

From the above let's review some of the statements made by some of our great minds regarding space. To the extent that Newton defined space we have no argument with space remaining similar and immovable when it is not in relation to anything external. When Einstein added the curvature of space we also have no argument since the concept of fields in the EST is a reorientation of the entropic quanta because of the presence of a mass, or a flow of charges (electricity and its associated magnetism). As to Mach's principle it is visualizable that the motion of one body would affect the motion and direction of another body in the vicinity.

The Einstein statement that there is no inertia relative to space (only masses react with masses) must be rejected by the entropic spacetime theory. There is no way to explain inertia as a mass-to-mass phenomenon except in making it a "property" of mass, without explanation. This is not particularly consequential to the relativity theory, since this statement which we rejected is not a pivotal point in the theory. It merely means that relatively was worked out without giving a basis for inertia.

CAUSALITY

We won't waste much space on this very profound philosophical premise. Newton, for example, disclaimed all hypotheses as to what gravitation really was. It was enough for him to describe it. Strict causal determinism with fixed values of measurable quantities have been given up in modern physics. Modern physics tells us that the world does not behave classically either in the microcosmos or on the cosmic scale. (Although it does behave classically in the middle region, the macroworld.) As a result there are no descriptions in familiar qualitative terms of the quantitative features of quantum physics, cosmology etc. The question is whether the nature of things is really indeterminate so that probability is ontological and not merely epistemological. Was the "Copenhagen school" really right or will further inquiry reveal that under the statistical surface of probabilities there is a determinate microstructure? If causality breaks down at the quantum level why are all electrons alike?

The entropic spacetime theory is also a quantum theory; its main premise is the quantization of charge, length and time. In the process of doing so we discover the universe is finite in all respects, filled with exact dimensions and furnishing the very opposite of what physics has been doing throughout the century; namely, making it impossible to explain anything except in terms of mathematics. Through EST we have a world of new causality. The elements of the EST allow epistemological treatment, meaning that we can seek their origin and nature.

GRAVITONS

If an astronomical body were to collapse or lose a great amount of energy, this would contribute a great amount of entropic energy to EST. This would induce a plane wave in the entropic spacetime which, like a wave from a pebble in a pond would traverse the universe. This effect should be observable in the deflection of the path of light if a clever observer could find a way of measuring it. As to actual particles, named gravitons, there is some doubt about their need to exist. The EST quanta actually do the job. Perhaps we should have renamed the scalar bosons—gravitons. But we chose not to!

THE DENSITY OF EST SPACE

The density of the EST space is crucial in the determination of the most exciting answers which this theory offers. One is the question whether there is sufficient energy in the universe to bring about cycling. Another is to explain HOW gravity works.

ENERGY CONTENT

It was stated in this theory that the universe is definitely cycling because the EST space provides the required energy many times over. The minimum density required, as stated by present authorities is a min. of 1×10^{-29} gcm^{-3}. EST is in S.I. units so we'll translate this into 8.98×10^{-22} Jm^{-3}. Let's assume this is correct.

How do we figure the EST density?

The ratio of E/ℓ is constant; it is the elastic constant k_{dil} which is given in units of Jm^{-1}. At first it would seem that the density of EST is constant. The third dimension, however is time and this dimension is variable.

TIME

We have several interpretations for time.

1) The time interval which is ℓ/c. Since ℓ varies with the expansion of the

universe so does the time interval. The time interval is a very small number. (See table of EST Values)

2) The time from the big bang until the equilibrium between the kinetic system and EST, which is the amplitude of the cycling universe or the time required to complete an epoch of the cycle. We call it t_T.

3) The time elapsed from the big bang until now which we will denote by t_E.

4) The time from now until the equilibrium which we denote by t_q. This time becomes smaller as we approach equilibrium and so it has a minus sign in front ($-t_q$). We call it Einstein time because in his field equations time is always preceded by a minus.

Which time shall we use?

To find the density at any time we multiply this constant of elasticity k_{dil} by the time interval. The basis of EST density is time alone

To find the EST density we use the dimension of time associated with the time-to-equilibrium t_q.

The EST density is $\rho = k_{dil} t_q$ and the units are Jsm^{-1}.

From the viewpoint of the kinetic system we think of density in units of Jm^{-3}.

The total energy content of space rises just as entropy rises. But the length ℓ also rises proportionately. Since the time-to-equilibrium, t_q, gets smaller as we approach equilibrium the net density decreases, while the total energy content of EST space increases. If this seems paradoxical, consider that space was more dense at the time of the big bang so it stands to reason that it becomes less dense as it expands.

In this theory, the cycle of the universe begins to reverse when the background radiation temperature is equal to the EST temperature. Entropy will have ceased and the stretched space quanta begin to shrink by gravity. (The gravitational length ℓ begins to shrink.)

We would want to know the density of EST space now and what it will be at equilibrium. (For ref. see Table 1)

The relative value of t_q to c (the universal constant) is 1 at the maximum density and at the moment of the big bang, while

$k_{dil} = 5.2224011 \times 10^{11} Jm^{-1}$. Therefore $\rho_{EST} = k_{dil} t_q = k_{dil}$

The relative value of t_q today is .5126 so that the density of space today is

$\rho_{now} = k_{dil} t_{qnow} = 2.677 \times 10^{11} Jsm^{-1}$.

At equilibrium the relative value of $t_q = \sim 0$. (We'll use the angle $\theta = 89.99999°$) so that at equilibrium

$\rho = k_{dil} t_q = 9.097 \times 10^4 Jsm^{-1}$

CONVERSION OF CONVENTIONAL DENSITY TO EST DENSITY

To compare these densities with a density given at Jm^{-3} we have to equalize the units for comparative purposes.

To do this we will multiply the kinetic density by $c^2 t_E^{\ 3}$, where t_E is the elapsed time relative value to c.

$$\frac{J}{m^3} \cdot \frac{m^2}{s^2} \cdot s^{-3} = Jsm^{-1}$$

Therefore 8.98×10^{-22} Jm^{-3} converts to 9.371×10^{-6} Jsm^{-1}. This is the proof that the density of EST space is far greater than the required density for cycling.

TABLE 1

ℓ as a fraction	time to equilibrium fraction of c = 1	Radial clock in degrees Θ	Elapsed time t_E fraction of c = 1
$\ell = \sin\theta$	$t_q = \cos\theta$		$t_E = 1 - \cos\theta$
~0	$t_q = 1$	0, (big bang)	$t_E = 0$
.85877	$t_q = .51236$	59.1787° (now)	$t_E = .48764$
1	$t_q = 0$	90° (equilibrium)	$t_E = 1$

(See CHAPTER V, The Inflationary Universe Theory, on more information concerning Table 1 data)

EST density, as described here is really ACTION PER METER; worthy of a moment of contemplation.

If $\rho_{EST} = k_{dil} t$ and $k_{dil} = E/\ell$ and $t = \ell/c$ then $\rho_{EST} = E/c$

E/c is the signature of momentum, so in the case of EST it would be potential momentum. This is the basis for "model making". A shrinking universe, releases energy from the potential to the kinetic state. It therefore induces the creation of radiation (vector bosons) which is thought of as momentum. In effect we have potential momentum as the "density" of EST and we have real momentum as the definition of radiation.

HOW DOES GRAVITY WORK?

Newton was contemplating how gravity works; he communicated with a friend that he thought it might be the difference in pressures in the ether, but being a meticulous scientist he never wrote it down for the world to remember.

Recall that in the treatment of density we show that the density of space is greater than the density required to bring about cycling. We also show that the density of EST is unique because it is given as an action per meter, rather than mass/volume which is the way we normally think of density. Let's differentiate between these two meanings of density: We'll call density which is m/volume kinetic—because it belongs to the kinetic system of physics, and EST density because it belongs to the EST system and has one dimension less than the kinetic system.

We also show that a particle which has the most density also contains less energy and, of course the converse, that a EST particle which contains more energy is necessarily less dense. This is because t_{eq}, the time from now to the equilibrium, gets smaller as we approach equilibrium, is the only variable. k_{dil}, the elastic constant represents the ratio of E/ℓ and does not vary throughout the cosmic cycle. Therefore EST gains energy and becomes less dense.

From observation and Newton's laws we know that energy content is the grav-itational element—the gravitational attraction of two space quanta is E^2/hc. The greater the energy value the greater the attractive force. The attractive force on any two bodies is $E_1 E_2 \ell_{now}^2/hc\ell^2$, where ℓ_{now} is the length of a space quantum today and ℓ (often called R) is the distance between the two bodies $E_1 E_2$.

BUOYANCY ANALOGY:

Let's try another approach now. On the subject of buoyancy, Archimedes said that a body is buoyed up by a force equal to the weight of the fluid displaced. Now, EST although not a fluid in the conventional sense, does get displaced by particles. A particle is a knot of wound up EST space and it displaces the scalar bosons of space. On earth, where we have a gravitational acceleration called g, the force of buoyancy is $F_1 - F_2 = \rho g V$ where ρ is density and V volume. In essence we're speaking of mg, mass and gravity as the basis of buoyancy. In any remote space EST does not sense any g. A single particle in remote space would feel no gravity. (Mach's principle) This is because the gravitational attraction of EST would be equal in all directions. Two particles or masses at a reasonable distance from each other would be attracted to each other by the mass (or E/c^2) of the displaced EST, in a principle very similar to buoyancy. When gas bubbles rise in a glass of fluid, we have a lower density of the bubble being attracted to the lower density of the air. This is a simplified analogy, because we omitted the g factor. But that is not important. The bubbles are also attracted to each other but the large area of the air overcomes the small mutual attraction of the bubbles.

When we attribute the gravitational attraction of bodies to their mass, it is not surprising. We've isolated the buoyancy to mg. In the absence of g it would seem that it is a matter strictly of the mass of the bodies. But we should not forget that the attraction is the force equal to the energy content of the EST displaced.

We've said nothing about the inverse square of the distance between the bodies. It is not the distance of the bodies alone, considering that we have shown the ratio of ℓ_{now}^2 / ℓ^2 which is really the controlling factor in distance. ℓ_{now} is an extremely small number so it would seem negligible but it is not. A very small number divided by a large number results in an even smaller number.

The essence of this analogy is that less dense bodies, embedded in a more dense medium, attract one another. When we talk of buoyancy we are really saying that the submerged bubble is attracted to the atmosphere which acts as the other low density body.

ANTIGRAVITY

One might ask what if a body was very very dense, more dense than EST space itself would that demonstrate reverse gravity? There is no particle of matter which is more dense than EST space. There is no length associated with any microbody which is smaller than the length of a EST quantum. That is why there is no antigravity.

This offers an additional reason for this theory's position that there is (or was) no singularity and why the existence of black holes is doubtful.

CHAPTER IV

COSMOLOGY

THE VAST MASS OF ENTROPIC SPACETIME

The widespread number of concepts pertaining to cosmology has been narrowed substantially. The most obvious deduction which can be drawn is that the entropic spacetime holds a vast amount of energy, as compared to the sum of all astronomical bodies and other visible or measurable forms of energy.

Findings from all astronomic sources suggest that even more than previously thought, the universe must consist of some *STRANGE UNSEEN MATTER*. The entropic spacetime theory strongly suggests that EST is that source of energy. Recent findings from the Keck observatory indicate that ordinary matter, the stuff stars and people are made of, is even more scarce than had previously been estimated. Most of the so called "dark matter" which astronomers have been searching for is invisible non-baryonic material. This theory says it is scalar bosonic energy and ubiquitous—not in the form of halos surrounding galaxies. (See *A RATIONALE FOR DARK MATTER*)

Are we to think that the scalar bosonic spacetime is uniformly populated with space quanta of such incredible density? Perhaps this is unlikely. Perhaps we should think of the densities as a gradient going out along the radius of the 'sphere universe' where the density of individual spacetime quanta increases— remembering that the higher the density the lower the unit energy content of the space quanta.

The idea of halos around galaxies stems from the uniform velocities of about 200 kms^{-1} of their outer branches. The uniformity of velocities takes place about 3.5 kiloparsecs away from the nucleus of the galaxy. This is distant enough to call the EST "remote space". The ℓ value of this space is the ℓ_{now} value. The nearer the space fabric is to the nucleus of the galaxy the higher the energy intake rate is, but that does not necessarily make it less dense. Astronomers are inclined to speak of halos because the effects are measurable at the outer perimeters of the galaxies. Actually, according to this theory remote space should be quite uniform in any local area. It is only the proximity of large energy sources which could alter the density of EST, but there is no real indication that this is so.

Energy is distributed through the fabric of space at velocity c; just like light. The density of EST, in the vastness of the universe could be more dense (contain less energy per quantum) in the most remote limits of the sphere-universe. At the risk of repetition, it should be noted that according to this theory, there is no other space than EST space. The sphere of the universe balloon is all the space there is. In this cosmology there is more space at the limit of expansion and less space at the limit of its contraction. Whatever the value of ℓ is, it is all the space there is.

ONE FRIEDMANN UNIVERSE

This theory has cast out most of the cosmological concepts offered to date except one: the so called Friedmann concepts. The general view by Friedmann was expressed in:

$k = H^2 (2q-1)$ where k is the curvature of space and H, the Hubble term. So a lot hangs on q, Friedmann's deceleration term. If q is greater than 1/2 then space is spherical and closed. If q is equal to 1/2 then space is flat and open but if q is < 1/2 the space is hyperbolic and open.

All Friedmann universes begin with a bang and are devoid of the Einstein Λ force. (The Λ force was what Einstein called his "greatest mistake" but it seems to be returning in many proposed cosmologies.) As we shall see as this theme develops, Λ is nothing more than the gravitational force of the entropic space itself which is constantly striving to be as small as possible.

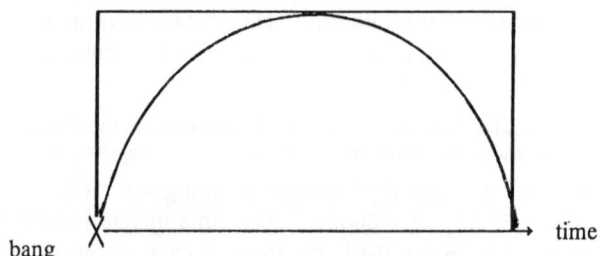

bang

time

FIG. 1 The Friedmann universe where q > 1/2

Why are we so convinced that only a closed cycle need be considered? Because, among cosmologists the question always hung on whether there was sufficient mass accounted for to provide the gravitational force necessary to cycle. The general opinion is that only 1% of the needed mass could be accounted for by direct observation. The entropic space provides a great deal more than a mere additional 99% of gravitational mass.

Having accepted a most likely concept, we can attempt to fit entropic space-time terms to the whole universe. So far we have values for the beginning of the entropic space. It contains a minimum energy just prior to the bigbang. The bulk of the total energy of the universe is divided between the bodies of matter and antimatter and the potential energy in the minimum universe-sphere. The EST quantum contains the minimum energy possible—h(1) per quantum, which also places it at the lowest temperature possible (but not absolute zero).

To restate the basic assumptions of the theory:

- The total amount of energy in the universe is constant throughout cosmic history, only its distribution varies.

- The total amount of charge in the universe is constant, only its distribution varies.

- All physical values in the universe are finite and have maximum and minimum values.

At its minimum, the entropic space quantum has the following individual values:

$$\text{Energy} = h(1) = \left[\frac{q_o^2 \, k_{dil}}{2}\right]^{1/2} = \frac{q_o^2}{2\ell_o} = E_o = 6.6260755 \times 10^{-34} \text{ J}$$

$$\text{Charge} = q_o = \left[\frac{2h^2}{k_{dil}}\right]^{1/2} = \left[E_o \, 2\ell_o\right]^{1/2} = 1.2966903 \times 10^{-39} \text{ C}$$

$$\text{Length} = \ell_o = \frac{h}{k_{dil}} = \frac{q_o^2}{2E_o} = 1.2687795 \times 10^{-45} \text{ m}$$

k_{dil} is the constant of dilation.

SINGULARITY

The current state of expansion of the fundamental space quantum described above, according to this theory is 3.1927028×10^{10} times the original value. How was this number arrived at? From the Planck length, from which we assumed the present value of $\ell_{now} = \ell_p.(2\pi)^{1/2}$. ($\ell_p$ is the Planck length. It seemed useful to change $\hbar$ to h in the scalar bosonic system. The 2π enters the picture when scalar bosons change their phase into vector bosons.)

Energy, charge and surface all increase as covariants. This multiplier was named ω. (In scalar bosons this is just the frequency expressing the stretch of entropic quanta but when these are converted to vector bosons then ω becomes the frequency.)

All things considered, the dilation number is not very large. With a great deal of energy now absorbed through entropy into space and starting from singularity this number should be much larger. The idea of singularity which has obsessed so many mathematicians and cosmologists, according to this theory, is impossible. If the total length of the sphere-universe has only increased some 32 billion times since the big bang, then the sphere of space at its minimum must have been quite large yet. This is not an unsupportable idea if we imagine the universe as a balloon which cycles from a minimum radius to a maximum radius and then back again. Within this balloon we have matter floating about which heats up as the balloon deflates and cools off as the balloon expands. This way the inviolate total energy is now in space and now in matter but never totally in one or the other.

This does not make the measurement of the universe an easy matter. For example the difference between ℓ_{now} and ℓ_p is only about 2.5 times. Had we accepted the Planck length ℓ_p as the present value of ℓ then the number of expansions since ℓ_o would be only 12 billion instead of 32 billion. This does not matter in resolving the question of singularity—it is too small a value for singularity in either case. But what about other measurements?

TEMPERATURES

We have certain limits which, according to this theory cannot be exceeded. If the EST is as low as it can be in the storing of energy, its energy is h(1) per quantum. The temperature also cannot be below a certain level (in this theory there is no absolute zero you will recall). Its temperature could not fall below:

$$T = \frac{h}{k_B} = 4.799216 \times 10^{-11} \text{ °K, where } k_B \text{ is the Boltzmann constant.}$$

If the EST is at its minimum temperature, is there a maximum temperature? To deduce this temperature we can draw on the fact that there is a maximum frequency ω_{max}. This in turn stems from the smallest time interval t_o

$$t_o = \frac{\ell_o}{c} = 4.2321928 \times 10^{-54} \text{ sec.}$$

$$\text{if } t_o = \frac{1}{\omega_{max}} \text{ then } \omega_{max} = 2.3628413 \times 10^{53} \text{ Hz}$$

Since $E = h\omega$, the temperature maximum could have been:

$$T = \frac{h\omega}{k_B} = 1.1339785 \times 10^{43} \text{ °K this being the highest temperature possible.}$$

Using the same approach we can also find the temperature of the present entropic quanta:

$$T_{now} = \frac{E_{now}}{k_B} = 1.5322469°K$$

This is an acceptable number because the average temperature of the kinetic stuff (cosmic background radiation) has been measured to be 2.735°K.

Any number less than that would be acceptable and greater than 2.735 would indicate an error in this theory.

The temperatures of the radiation background in the kinetic system and the EST appear to be quite near each other suggesting that a great deal of the overall energy supply has already been deposited in the entropic spacetime. (The other stations for the potential energy are the rest masses of the particles which make up the universe.) The rest masses are locked up in stable states—protons, electrons, neutrinos and photons. The hot galaxies which are melting away by radiation should contribute energy to the EST.

We can roughly estimate the equilibrium temperature : 1.785°K

After equilibrium is achieved there will be no differential between cosmic radiation temperature and the temperature of the entropic space. Entropy in space will cease. The main transfer of energy in space will cease, but at higher temperature differentials the second law of thermodynamics may still operate. At the macrolevel at which we live the equilibrium should hardly be observed. Only the ultimate heat sink (entropic spacetime) will stop working.

EQUILIBRIUM DATA ON SPACE QUANTA

The approximate values for the space quanta at 1.785°K should be as follows:

$$E_{eq} = Tk = 1.785(1.38 \times 10)^{-23} = 2.463 \times 10^{-23} J$$

$$\ell_{eq} = \frac{E}{k_{dil}} = 4.717 \times 10^{-35} \, Jm^{-1}$$

$$q_{eq} = \left[\frac{2E^2}{k_{dil}}\right]^{1/2} = 4.8299 \times 10^{-29} \, c$$

How long will it be before this happens? This brings us to the problem of the age of the universe.

THE AGE PROBLEM

The age of the universe, according to the entropic spacetime theory, cannot be converted into meaningful years. The present method of measuring time by years is not satisfactory. The 15-20 Billion years, the current estimate or the 2 Billion-Hubble's original number are apparently meaningless. Why?

Consider the following:

A time interval is applicable to the state of the quantum at a particular condition of expansion:

At the beginning $t_o = \dfrac{\ell_o}{c} = 4.232192910^{-54}$ sec

$$t_{now} = \dfrac{\ell_{now}}{c} = 1.3512134 \times 10^{-43}$$

$$t_{eq} = \dfrac{\ell_{max}}{c} = 1.573 \times 10^{-43}$$

$\dfrac{t_{now}}{t_o} = 3.1927028 \times 10^{10}$ or ω_{now}

In effect, at the beginning we've had a year which was the equivalent of small fraction of a second today. The age of the universe made up of these tiny, but growing years, seems meaningless.

As far as the universe is concerned we can think in terms of number of times it was multiplied (by stretching). In those terms we can say that the current age is 31.927 Billion times.

At equilibrium it should have a frequency number of 37.17 billion times.

The entire history of the entropic spacetime is predicted to exist in temperatures below the cryogenic λ point. (The maximum temperature of the EST at equilibrium is predicted to be in the vicinity of $1.785°K$.)

AFTER THE ENTROPIC EQUILIBRIUM?

The entropic spacetime begins to shrink due to what we call today gravitational pull. The real meaning of that is that the time interval becomes smaller and along with it all entropic spacetime values. The process should be as slow contracting as it was expanding. Some cosmologists have a way of referring to that possibility as "collapsing" space, insinuating a suddenness to that process. We expect symmetry on the curve. A good deal of matter is expected to remain intact at the point of equilibrium.

The reverse process means that the quanta will decrease in size and yield energy.

This means that the remaining matter will increase in kinetics in a process where scalar bosons will become the building blocks of everything—the opposite of being the heat sink for everything.

Increasing kinetic energy of matter will bring about an increase in collisions and interaction. This interaction should result in the creation of new matter from the surrendered energy of the contracting space. The new bias of the universe will be that the new matter created will be antimatter. As the contraction increases more and more, antimatter is created in an increasingly kinetic universe which is getting smaller. As we have already stated previously the shrinking process should not at all go to a "singularity" but would instead be merely decreased to a smaller sphere.

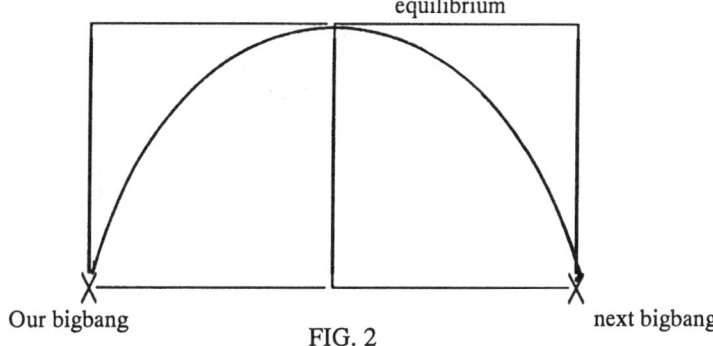

equilibrium

Our bigbang

FIG. 2

next bigbang

It is the accumulation of new antimatter which should lead to the next bang, rather than the concentration of all the energy into a singularity, as is speculated today. The next expanding cycle should show a universe in which only antimatter will be around. This, of course, provides a reasonable explanation why we lack a natural supply of antimatter in our epoch. It also provides a reasonable explanation for the big bang in the first place. The potential energy in particles, as rest energy and the potential energy of the entropic space, become the kinetic energy of the next epoch, thus fueling the continuity of cycles.

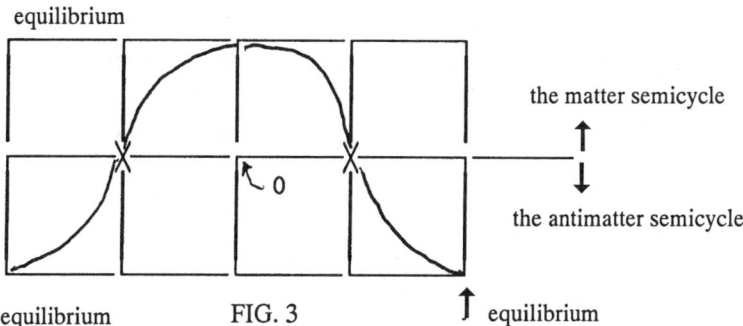

equilibrium

the matter semicycle

the antimatter semicycle

equilibrium

FIG. 3

equilibrium

AT EQUILIBRIUM:

The mirror image is the same as if the curve pivoted 90° about point O. The time and space axis are now in a changed orientation; time becomes a horizontal component and space is vertical. The reversal of t and s would seem to have little significance except that it may provide a reversal of B and E in electrical and magnetic fields. Perhaps this is the underlying basis of matter and antimatter. The reversal of the discernible charges is easy to see.

In the contracting epoch the new matter created should abide by the left hand rule—just as now, we have the right hand rule to show the flow of current and direction of magnetic field.

Back to the condition of entropic spacetime: The reduction in size of ℓ means reduction in quanta energy, reduction in the electric and gravitational force in the quanta and reduction in temperature of space. The kinetics of the old matter and the new antimatter are increasing, with a temperature increase there. The energy is draining out of the entropic into the kinetic state.

Is the second law of thermodynamics invalid in a contracting universe? The answer seems to be definitely yes, as far as the interface of the kinetic and entropic space are concerned.

However we may be in the n'th cycle, with, who knows how many times, this has occurred already. It is entirely possible that in the beginning there was a singularity where each repeating cycle obeys the second law of thermodynamics and the maximum and minimum dimensions of a spherical-universe are approaching each other at each repeating cycle. In other words, are the cycles obeying a grand law of entropy and are becoming damped out?

TOUCHING ON QUANTUM FIELD THEORY

In quantum field theory, as far as it is known today, all elementary interactions are based on exchanging field quanta. At first this seems to be contrary to this theory. But is it? Let's review the entropic spacetime concept.

Field quanta are bosons. In an expanding universe the creation of new matter is not possible. In a contracting universe (the next cosmological epoch predicted in this theory) vector bosons and matter are created directly from the scalar boson field, where entropy is reversed.

In the present expanding universe, fundamental interaction is created from the downward cascading of larger particles into more fundamental particles. Energy is added to the entropic spacetime. In this expanding universe exchanges with entropic spacetime occur all the time—as with so called virtual particles. But what are vector bosons and particles but kinetic versions of the same stuff that scalar bosons are made of. When it is in the scalar state it is spacetime (plus energy and charge). When it is a vector boson (photon) then it is the same stuff as a torsion wave which gives it the property of momentum. When it is a fermion then it is the same stuff only with angular momentum which gives it the property of inertia. During the expansion state the energy of the kinetics comes from the breakdown of matter and the absorption of vector bosons. In the contracting state, the kinetics come from the field of scalar bosons. The interaction is the same— the source of the kinetic energy is different. In the first case it is decay, in the second it is the result of the action of gravity.

The entropic spacetime, in quantum mechanics terms is in spin 0. Vector bosons are in spin 1. Fermions are in spin 1/2 . In quantum mechanics it is said that vector bosons (photons, gluons, W$\pm$ and Z° particles) are intermediates in these interactions. But they are more than that too. They, the intermediates, are the building blocks themselves as energized scalar bosons.

The process of exchange can occur by the same topological process in a decaying universe with the source of energy coming from the decay as it does in a contracting universe where the source is the bosonic (spin zero) spacetime.

It is commonly thought today, that gravity belongs in a spin 2 quantum environment—probably following Feynman's argument on this subject. He showed that in a kinetic environment, spin 0 would not fit observations on gravity, but it would rather show the inverse of the properties of gravity. We are arguing that the properties of entropic spacetime are inverse to the properties of the kinetic system. The inverse character of entropic space vs. kinetic space makes spin 0 exactly right. Inversion, or the placement of a minus sign in front of an exponential also explains Dirac's thinking on negative energy. Entropic spacetime is negative vs the kinetic system in regard to being the storehouse of potential energy in an expanding space epoch. It is not necessary that this potential energy be negative in some mystic way. It is simply energy which is subject to the bias of an expanding universe.

ABOUT MATTER COLLAPSING INTO HIGH DENSITIES

Somehow, there has been the general belief that matter can be collapsed to far greater densities through some cosmic process. Here we have argued that space particles are far more dense than particles of matter. Since mass/energy is the qualification for gravitational force rather than density, we see that a proton indeed does have greater gravitational force than the electron, but not because of its density but rather because of its energy content. It was never shown that a nucleon can be made more dense than it is. A mass can be increased by velocity, but that does not make it more dense. The densest stuff we have in the entropic space theory is the fundamental quantum, which contains the lowest energy level.

The idea of superdense material made up of nucleons packed without limit has never been demonstrated, or found or observed. In fact the more nucleons are packed in an atom the less stable the atom becomes, with the vicinity of 300 being the limit in the heaviest atoms. We can only conclude that within the limits of this theory, there is no superdense supergravitationally powerful singularity possible.

THE END OF THE CONTRACTION EPOCH

Once a conceptual system is described, it isn't right to leave loose ends. We now can read detailed descriptions, second by second of the beginning of the universe; now we'll dwell for a bit through the moments prior to the big bang.

The entropic space has been contracting as previously mentioned. There has been quite a bit debris and matter around. The energy in the entropic space is quite low compared to the vast amount which was stored there at the equilibrium. This energy, released, has created antimatter. Much of this creation has conglomerated into giant bodies of new antimatter. The matter itself has bunched up into giant bodies. Sooner or later the two meet and the conflagration occurs. There is less matter than antimatter so the new epoch comes into being with a bang and the resulting and remaining material is now antimatter. There is no law that says that the big bang has to happen all at once. The bodies of matter and antimatter can find each other over a period of time. The main thing is that they will annihilate each other and expand the universe in the process.

One half of the cycle of the universe, starting from our own big bang has been completed.

PEERING TO THE EDGE OF TIME

A recent discovery which has upset theories of cosmic evolution, was the most distant light from quasars. Analysis of the light emitted by the newly discovered quasar indicated that it existed when the universe was only 7% of its current age. What makes this discovery startling? It indicates that galactic sized objects must have formed when the universe was very young. Essentially, there is too little time for the creation of galaxies, if we adhere to the present cosmological schedules.

A team of Caltech astronomers, led by Dr. A.C.S. Readhead, said that their new findings included evidence suggesting there could be enormous amounts of matter in the universe that is fundamentally different from ordinary known matter. This missing mass must be central to the explanation of how matter came together into its present state. This is precisely where the character of entropic spacetime comes in.

Another recent revelation stems from the Cosmic Background Explorer (COBE), an unmanned craft designed to yield measurements of our early existence which could explain the origin of the galaxy clusters. From early data from COBE, Dr. George Smoot, University of California at Berkeley stated: "Using forces we know now, you can't make the universe we now know." It only confirms the recent evidence from the most distant of quasars that there is no time in the present cosmic schedule to create galaxies. If we follow the cosmology suggested by the entropic space theory, we have no singularity, merely a minimum space sphere at the time of the big bang. The matter we know now is that which has survived the matter-antimatter annihilation of the big bang. It was there all along and we need no period for the creation of galaxies as the present concepts require.

Immediately after the big bang we had a great deal less of matter than in the epoch preceding that cataclysmic event. Most of the matter and antimatter annihilated each other and provided the fuel for the expansion of the universe. The matter in the cosmos now is the surplus of that annihilation. This would account for the smoothness of the early post-big bang period as well as its present condition.

MORE ON WHEN THE GALAXIES WERE CREATED?

The most distant object recently found at the edge of the universe is a gigantic gas cloud of carbon monoxide. The distance is estimated by the discoverers R.L. Brown and Paul Vanden Bout as 12 billion light years. The assumption is that this is a galaxy in the formation process. Its age is assumed to be 3 billion years after the big bang. Are we to conclude that this formation of gas has occurred that soon after the big bang? If carbon and oxygen are produced only in stars and not directly in the big bang (due to the explosion of a singularity) then there must have been at least one generation of stars which have already completed their life cycle. When could this have been? The previously suspected flaw in the schedule of post big bang events is getting worse from this discovery. It would confirm the following:

1) That the big bang is only one of a series of such cycling events.

2) That our big bang is the result of a compression phase which brought together matter and antimatter of the previous epoch.

3) That the present expansion epoch already contained all the elements—rather than have them formed from galactic action in our epoch.

4) This line of thought gives the concept of singularity a tremendous blow. If it occurred at all it was not in our epoch but n cycles ago.

A RATIONALE FOR DARK MATTER

In our Solar System planets revolve around the sun at different velocities. The closest, Mercury at about 48 kms^{-1} and the furthest, Pluto, at about 5 kms^{-1}. This is in perfect agreement with Newtonian calculations, where the sun in the gravitational center and the planets whirl around in a vacuum-continuum.

However the rotational velocity for clusters of galaxies differs greatly from this. As far back as 1933, Zwicki noted that individual galaxy clusters should fly apart unless another large gravitational mass were present to hold them together.

40

Hence the mystery of Dark Matter.

The detailed studies since that time indicate that the orbital velocities act just the opposite of our solar system, and then level off at about 200 kms⁻¹. The conclusion that there must be dark matter is inescapable. The astronomical observations confirm the existence of large amount of energy, other than that present in the central portion of the galaxy. The speculation is that this energy extends well beyond the limits of the optical galaxy—but without a specific suggested distance.

On Fig. 1 below we see the velocity range in the solar system.

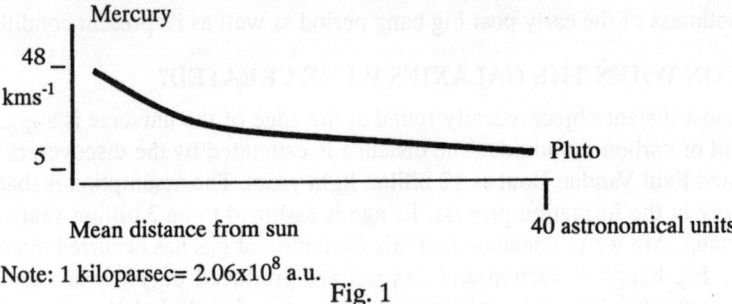

Note: 1 kiloparsec= 2.06x10⁸ a.u.

Fig. 1

ORBITAL VELOCITY OF THE SOLAR SYSTEM

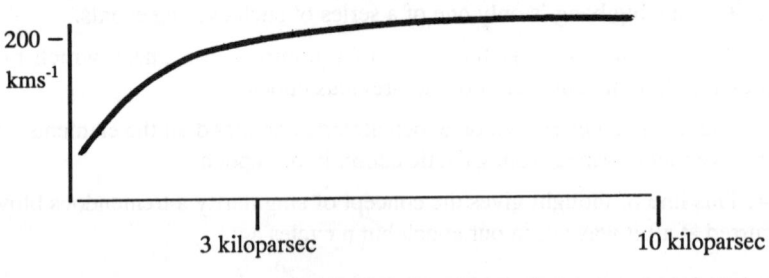

Fig. 2

ORBITAL VELOCITY IN TYPICAL GALAXY

The difference in scale between Fig.1 and Fig. 2 is so great that the distance in the first is completely lost in the second. In Fig. 2 we see the gravitational effect of the "dark matter". Orbital velocities rise in the first 3 kpc and thereafter it levels off. Can these observations be used to support the entropic spacetime theory?

A REVIEW OF THE ENTROPIC SPACETIME THEORY

We would have to review the highlights of the theory to see if it does indeed offer an explanation. After all, theories abound; Is it axions, neutrinos, weakly interacting massive particles (wimps), the magnetic poles of Dirac? Nothing in the Standard Model offers a clue. The simplest answer seems to be the existence of another fundamental particle—a scalar boson (with a quantum spin value of 0) different from the vector boson with a spin of 1. Scalar means to have a numerical value without the momentum of the vector bosons (photons, radiation, light etc.)

In order to be a factor in gravity at all it, the scalar boson must have energy.

To be a scalar boson—one without momentum the energy cannot be E/c—the signature of momentum.

In order to function ubiquitously it must be space itself and therefore cannot be detected directly. (As Dirac said: a fish cannot detect the presence of water.)

It could not be a fermion ($m = E/c^2$) because all mass particles can be detected.

To be space, it must have the properties of space. μ_o, the permeability of vacuum and ϵ_o, the permissivity of vacuum and c must play a role in the definition of space.

$$\frac{E}{\mu_o \epsilon_o c^2} = \frac{E}{1} = E$$

Let's describe space: A ubiquitous field of pure energy in quantum form. Each quantum is electrical in nature and spread in surfaces. ϵ_o requires that an increase in capacitance requires an increase in energy and distance. The electrical neutrality of space requires that the charges on the capacitor surfaces are of opposite types. The stretching of the distance between the surfaces of the capacitor requires energy.

The first law of thermodynamics requires that energy is conserved.

The second law indicates indirectly that energy flows from higher levels to lower levels. At the same time the law indicates indirectly that entropy always rises. The stretching of space then means that space absorbs energy from its surroundings in order to conserve it. It also increases in volume as entropy increases.

All things subject to gravity must contain energy. If space contains energy, according to the logic above, it also provides a non-central source of gravitational energy.

INTERNAL GRAVITATIONAL FORCE OF SPACE QUANTA

The gravitational force between any pair of space quanta works out to be:

$$F_G = \frac{E^2}{hc}$$ this can be derived from the Newtonian eq. below

$$F_G = \frac{Gm^2}{R^2}$$

Substitute E/c^2 for m. From Planck's length eq.:

$$G = \frac{\ell^2 c^3}{h},$$ ℓ and R are both lengths

$$F_G = \frac{\ell^2 c^3 E^2}{h \, \ell^2 c^4} = \frac{E^2}{hc}$$

This is based on the supposition that remote space has a length of $\ell_p . (2\pi)^{1/2}$, Planck length multiplied by the square root of 2π. This length is assumed to be the size of the quantum today—stretched from an earlier size when it contained less energy.

ON GREAT ATTRACTORS, THE WALL and THE GREAT VOID

The concept of the uniformity of the universe was shattered when in 1987, seven astrophysicists announced that every nearby galaxy was streaming @ 600-700 kms[-1] toward a direction, some 300 million light years beyond Hydra Centaurus. After two more years of search the Great Attractor was named. The gravitational effect was some 20,000 times greater than the Milky Way. Then to the astonishment of the theorists the velocities of the galaxies ceased altogether some 150 million light years out. Apparently they were now at the center of attraction. It should be noted the frontal velocity of a group of galaxies is about π times larger than the typical rotating velocity described earlier (about 200 kms[-1]). This supports the idea of universal presence of EST. It also supports the idea that the density of EST may not be uniform in the vastness of the universe, notwithstanding the velocity c which is used to distribute unevenness in the density. Energy is constantly fed to the EST through the entropy process. It differs from the uniformity of the radiation background resulting from the cooling of the great explosions (big bang). (See *THE INFLATION THEORY...*)

Then other attractions were found—exactly in the opposite direction of the Great Attractor. From this we can only conclude a vast gravitational network being stretched and creating a universe in tension. Instead of a uniform distribution of galaxies, astronomers found great clusters of galaxies and supergalaxies called attractors. Even more surprising was the discovery of great stretches of "empty" space. These voids were estimated at 300 million light years across.

In a feat of mapmaking by Huchra and Geller galaxies were arranged in "bubbles" 150 million light years wide with empty interiors. Then came the discovery of the Great Wall made up of galactic bubbles. There were indications that the Great Wall might be one of a series of gigantic sheets lined up in a honeycomb structure with voids some 400 million light years apart.

One should credit Vera Rubin for her foresight, when she completed a paper in 1954 pointing out that galaxies were clumped with enormous spaces in between; nobody seemed to pay attention then!

It should also be made clear that when these irregularities are quite independent of the overall expansion of space.

Does all this recent astronomical data fit into the entropic spacetime theory?

Where vast voids are found—imagine vast energy fields—with the highest densities. (Essentially space that has been least stretched). The decaying energy of the galaxies are slowly feeding energy into this tight network of quanta—enlarging them and making them less dense.

It should be emphasized that when we call entropic spacetime SPACE, there is no other vacuum or setting for entropic spacetime to exist in. There is no blank page upon which the idea of the universe can be drawn! According to this theory there is no continuum.

This theory calls space as the creator of all gravitational phenomena—while masses are only the pawns subject to the forces of space.

This theory adds Energycharge to the Spacetime geometry and creates a quantized , finite and discrete universe capable of delivering deterministic values without necessarily disturbing the probabilistic results found in quantum mechanics.

THE UPPER LIMIT

Any theory which is based on a finite universe must have not only a low limit of dimensions but an upper limit as well. For example, the smallest wavelength possible denotes the largest frequency. The largest frequency also indicates the largest energy content. The smallest time interval t_o provides the basis for the largest frequency. ($\omega = t_o^{-1}$)

The following are the maximum values. No particle, boson or fermion can have values higher than the values given below:

$$\omega_{max} = t_o^{-1} = 2.3628413 \times 10^{53} \text{Hz} \quad \lambda = 1.2687795 \times 10^{-45} \text{m}$$

$$E_{max} = \hbar\omega = 1.5656365 \times 10^{20} \text{J} \quad \ell_{max} = \ell_o \omega_{max} = 2.9979246 \times 10^8 \text{m}$$

$$q_{max} = (2\ell E)^{1/2} = 3.0638734 \times 10^{14} C \qquad y_{max} = \frac{\ell E}{hc} = 2.3628413 \times 10^{53}$$

$$T_{max} = \frac{E}{k_B} = 1.1339785 \times 10^{43} \,^{\circ}K$$

THE DIMENSIONS OF THE UNIVERSE

We attempted to arrive at the dimensions of the universe from EST data. The following is a step-by-step method used:

The volume of an EST quantum would be $V = \ell t$. The units would be in ms (meter seconds). The potential nature of all dimensions of space create this situation. We need to convert this volume into cubic meters. By multiplying the potential volume by ℓc we obtain m^3. (see Table of EST Values)

If $V = \ell^2 tc$ and $tc = \ell$ we have $V = \ell^3$. This seems obvious once you see it but for a length ℓ which is made up of two dimensions $\ell = \lambda y$ this is not so obvious. True that y is a unitless dimension so that ℓ is in meters—a conventional volume should be ℓ^3.

$$\text{Conventional Volume} \quad V = \ell_o^3 \quad = 2.042483 \times 10^{-135} m^3$$
$$V = \ell_{now}^3 = 6.6469 \times 10^{-104} m^3$$
$$V = \ell_{eq}^3 = 1.0494 \times 10^{-103} m^3$$

FREQUENCY ω

This frequency exists in the EST as potential frequency and is a count starting from 1 (meaning $1h$ when we consider energy). The scalar bosons of EST have a high limit for the potential frequency at about 4×10^{10}. This potential frequency lacks the dimension of t, it is just a number which denotes dilation of the EST. This amazing number also counts the length ℓ and the charge q of the EST and the potential frequency also reflects the dilation of E, q, ℓ of the same vintage. We should be able to work out the energy content in a cubic meter as well.

In the kinetic system this frequency denotes Hz or cycles per second. Once we leave the world of scalar bosons and enter the kinetic world of vector bosons we have the frequencies, again from one to some maximum number. How do we know there is a maximum number? It is because in a discrete universe there is no dimension less than ℓ_o. That means we cannot have a wavelength shorter than ℓ_o. Since $c/\lambda = \omega$ the smallest possible wavelength will give us the highest frequency possible. We don't know if there is a thing which has such a frequency but

$$\omega_{max} = 2.3628413 \times 10^{53} Hz\text{—The Cardinal number of the universe.}$$

(For those who are fascinated with numerology, including Eddington, Dirac, Noyes and many others, who see the relation of natural unitless numbers as some special magic, will be disappointed in this number. Entropic spacetime theory has no interest in numerology.)

From the volumes of the space quanta given above, the ratio, of say, $V_{eq}/V_o = \omega_{eq}{}^3$. The proportionality of the volumes vary as the cube of the appropriate frequency ω. The volumes are given in cubic meters. The cube of ω_{max} might just help us find the volume of the universe.

To sum these values up, we have:

$\omega_{max}{}^3(\omega_o{}^3)V_o = 2.6944003 \times 10^{25} m^3$ (smallest volume)

$\omega_{now}{}^3(\omega_{now}{}^3)V_{now} = 2.8537314 \times 10^{88}\, m^3$ (present volume)

$\omega_{max}{}^3(\omega_{eq}{}^3)V_{eq} = 7.1105 \times 10^{88}\, m^3$ (volume at equilibrium)

Allowing for the fact that this method of establishing the size of the universe is somewhat speculative, it must be very clear, nonetheless that the smallest volume is far, far greater than any singularity. This is one of the cornerstones of this theory.

We can now work out the radii of the universe when it is at it's smallest and when it is as large as its going to be before it starts to shrink.

$$R= \left[\frac{3V}{4\pi}\right]^{1/3} = \left[6.4324069 \times 10^{24}\right]^{1/3} = 1.859764 \times 10^8 m \text{ (smallest radius)}$$

$R_{now} = 1.8957227 \times 10^{29} m$ (the present radius of the universe)

$R_{eq} = 2.5700 \times 10^{29}$ m (the maximum radius of the universe at equilibrium)

The furthest sighting from any of our astronomical instruments have been about 10^{12}-10^{13} lt yrs. A light year being 2.5062×10^{15} m we can say that with our best instruments we are now peering at the edge of time. We can also say that in the next 15-20 billion years we will not see much further.

Our present limit from this calculation is 7.5641318×10^{13} lt. yrs. The furthest we will ever be able to see is 1.0254569×10^{14} lt. years.

With values given above we should satisfy one's curiosity to see what the energy content of the EST would be.

The quanta per cubic meter at the minimum and maximum extension of the universe:

quanta at beginning $V_o^{-1} = 4.896 \times 10^{134} m^{-3}$

quanta at equilibrium $V_{eq}^{-1} = 9.5292 \times 10^{102} m^{-3}$

Energy per quantum, (from Tables)

$E_o = 6.6261 \times 10^{-34}$ J $= 4.1357 \times 10^{-21}$ MeV

$E_{eq} = 2.436 \times 10^{-23}$ J $= 1.537 \times 10^{-10}$ MeV

Volume of the universe:

$V_{Uo} = 2.6944 \times 10^{25}$ m^3

$V_{Ueq} = 7.1105 \times 10^{88}$ m^3

Total energy in EST: .

$V_o^{-1} E_o V_{Uo} = 5.456 \times 10^{139}$ MeV

$V_{eq}^{-1} E_q V_{Ueq} = 1.0414 \times 10^{182}$ MeV

The estimated amount of matter (by others) in the visible kinetic system is about 10^{78} MeV.

RECONCILING METHODS

In Chapter III, Gravity, we dealt with density; we compared a minimum required energy density for a cycling cosmology with the EST density. This was done by a different method than that shown on the TABLES. We did not use the time interval ℓ/c which is used here to determine the volume and density of the space quantum. Instead we used a cosmological time t_{eq}, the time to equilibrium to establish that space density is far above the required minimum. The t_{eq} is not an actual value but a $\cos \theta$ of a universal constant.

In both methods we proved the immense density of space.

CHAPTER V

THE INFLATIONARY UNIVERSE THEORY IS IT NECESSARY?

THE HORIZON PROBLEM

The large scale uniformity observed in the universe seems to defy the fact that the distances at opposite end of our visible sky are greater than the velocity of light could cover. In other words the standard model universe evolved too quickly.

At any given time there is a maximum distance known as the horizon distance. These horizon distances, when placed in a light cone, not only don't overlap, but are far apart from each other. This indicated that no physical process could have brought about the smoothness and uniformity which is observed. At least not at velocity c. In fact, in the standard model the sources of background radiation, when observed from opposite directions in the sky, were separated more than 90 times the horizon distance. Since the regions could not have communicated how can conditions be so nearly identical?

The Inflationary Model used a "false vacuum" phenomenon to accelerate the expansion of space beyond the time increase. To make this theory work it was necessary to call on two Higgs fields, some true vacuum states and...

UNLESS...

Unless the initial condition, meaning the nature of the big bang, was particularly smooth, so it was not necessary to communicate equalization between the regions. That is all very well but if that is the case when did the clumping of matter occur to create the beginning of the galaxies?

We'll try another way to account for these mysteries by way of the EST theory.

EST places new concepts on the meaning of time and space. The theory is based on the ubiquitous presence of energy, which is defined as space. Space is

48

therefore quantized and consequently the dimensions of these quanta, time and space are also quantized. The expansion of space is caused by the flow of entropy from the general decay of all suns and galaxies in addition to the initial engine—the big bang. The vast amount of energy which is space itself, guarantees a cyclic cosmology, and satisfies the need for "dark matter".

EST also states that there is no singularity, as visualized by the interpretation of Einstein field equations. From the deterministic aspect of the theory it was calculated that the universe will expand only some 37 billion times from big bang to the equilibrium. Our current state is that we have already expanded some 32 billion times. From the above, we visualize a universe balloon which after reaching a maximum diameter will shrink to a balloon of still enormous size. Hence—no singularity.

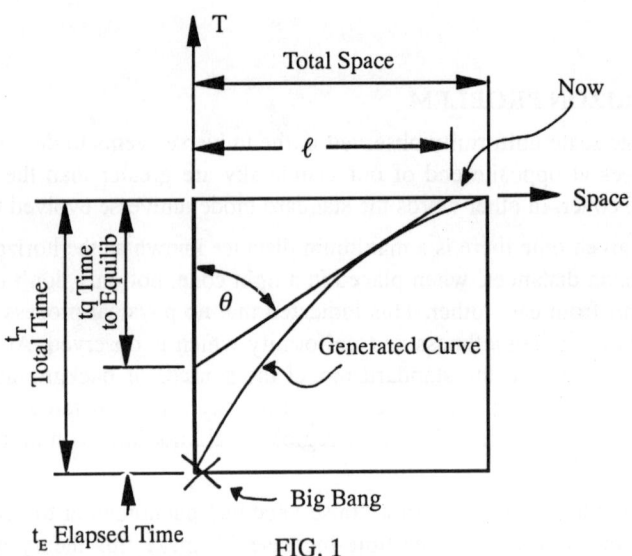

FIG. 1

If we think of time and space as coordinates we wish to develop a curve which will represent the expansion of the universe. Time and space have one constant common to both, c. In the diagram FIG 1. we will have c representing the total time t_T from big bang to equilibrium and the ℓ_{max}, the maximum space expansion for that same period.

We note that in the t_T there are two components: one is the ever diminishing t_q the time remaining until equilibrium and t_E the elapsed time from the big bang. When we think of the age of the universe we think of t_E, the elapsed time from big bang.

The constant c which generates the curve in Fig 1. is always in contact with the time and space coordinate. The angle θ gives us the spacetime measure from 0-90° degrees, denoting both time and space from big bang to equilibrium. Space, $\ell = \sin \theta$

$t_q = \cos \theta$ (time to equilibrium)

$t_E = 1 - \cos \theta$ (time elapsed)

It can be seen that when θ is small we have small time elapsed vs large space expansion. For example take $\theta = .1°$

The ratio of ℓ/t_E is 11,635 times. In the EST theory we don't need an Inflation theory—it is built in.

As θ approaches 90°, the ratio is one.

The Inflation theory is used to justify its need because there has been little or no sign for the beginning of the building of galaxies in the standard model physics. Not that the Inflation theory gives us an answer to that question but at least it explains the uniformity.

The EST theory is very clear in stating when galaxies are built. It is in the epoch prior to an expansion epoch. In the compression epoch energy is released from the EST which is capable of being the building block of both radiation and matter. This is described in detail in the model building. The big bang is according to the EST theory a matter-antimatter explosion. If the compression epoch builds antimatter it collides ultimately with the remains of the decaying galaxies of the previous epoch. The next expansion epoch would therefore be an antimatter world just as ours is a world of matter.

The theory also states that no new matter is made in an expansion epoch such as we now live in. Yes, we combine existing nucleons into new elements and molecules. No example of new hydrogen by natural means has ever been witnessed or proved. All we witness is the breakdown of large particles into smaller ones in a cascade downwards.

The idea of a singularity exploding into hot gasses which in turn cool and create matter is according to this theory false. That is why ALL IS STILL as we look backwards into the radiation background. An exploding galaxy or star somewhere out there, does not prove the creation of single proton which was not there already. All we have noted is the release of nucleons from more complex elements or the fusion of nucleons into heavier elements—both with the release of energy. In no case was there the creation of a new nucleon. This argument is of prime importance in support of the EST theory.

Miscellaneous observations:

A curious matter is the fact that in all of Einstein's equations time is negative. This has been the subject of considerable discussion—in some cases it was used to attempt to discredit relativity altogether. In our case the negative sign for time is a time interval on t_q, which is ever diminishing.

There is also curious matter concerning time. When Planck wrote ℓ/c as being time, there is a question of whether he meant $\ell/c = \sin\theta$ or the time to equilibrium t_q (Einstein time). If Planck meant ℓ/c then the time is total time—from big bang to equilibrium. Since there was no cosmic concept of equilibrium at the time Planck wrote his mysterious three equations we could only assume the $\sin\theta$ interpretation.

The curve generated by the constant c creates the Lorentz transformation. In relativity the transformation is applied to kinetic situations: velocity/c, momentum, kinetic energy etc. In our case the curve sheds light on relativistic relationships in the scalar universe of EST.

The relation of space ℓ to t_q, the time-to-equilibrium (Einstein time)

$$\ell = t_q \cos\theta = t_q \left[1 - \sin^2\theta \right]^{1/2} = t_q \left[1 - \left[\frac{\ell}{c} \right]^2 \right]^{1/2}$$

The relation of elapsed time to space ℓ:

$$t_E = 1 - \cos\theta = 1 - \left[1 - \sin^2\theta \right]^{1/2} = c - \left[1 - \left[\frac{\ell}{c} \right]^2 \right]^{1/2}$$

The relation t_q, time-to-equilibrium to space ℓ:

$$t_q = 1 - \frac{\ell}{\cos\theta} = \frac{\ell}{[1-\sin^2]^{1/2}} = \frac{\ell}{\left[1 - \left[\frac{\ell}{c} \right]^2 \right]^{1/2}}$$

WHERE ARE WE TODAY?

Let's call the curve which yields the above relations the sin-cos curve or the Lorents curve (L. Kauffman, the knot theorist, named it the ladder-on-the wall curve). It gives us the relation of space length ℓ to two kinds of time; the elapsed time from the big bang and the time to the equilibrium. Where are we on that curve?

How much has the universe expanded and how much time have we used to get there? The values of ℓ are given below for reference. How we got these values is described elsewhere.

$\ell_o = 1.2687795 \times 10^{-45}$m (the smallest value of ℓ)

$\ell_{now} = 4.0508359 \times 10^{-35}$m (the present value of ℓ)

$\ell_q = 4.717 \times 10^{-35}$m (the value of ℓ at equilibrium)

From these values we see that space has expanded 85.877% of the amount it will reach at equilibrium. Time however is another matter.

t_q = 51.2355% (time-to-equilibrium still has about half of the total time to go)

t_E = 48.7655% (time elapsed of the total time)

From this we see that most of the expansion has already taken place but we have used only half the time to do it.

GRAVITATIONAL LENS BENDING

Photographs from the Hubble Space Telescope have detected a striking example of gravitational lens bending. Light from a distant galaxy has been focused and concentrated into images which are bigger and brighter than they would otherwise be. This lensing effect has been attributed to *dark matter*, a mysterious and as yet undetermined source of energy. Now this lensing effect is a function of a galaxy cluster's gravitational force.

In this theory there would seem to be a clear explanation of this effect. Large bodies have enormous dimensional EST values as they are embedded in the entropic spacetime. Light traveling through space follows the shortest time line. This was predicted long ago by Hamilton-Fermat: "Light passes from one point to another in a way which minimizes the time of passage". Einstein concurred with this thought. Now, we are saying that light travels along lines of the entropic spacetime. This would lead one to think that the so-called Dark Matter is the entropic spacetime as described in this theory. When this curvature resembles the action of a lens then we have gravitational lens bending. The black hole which is often attributed to exist in the creation of such curvature need not be present to bend spacetime.

TIME

The meaning of time has been the subject (and even the obsession) of every kind of treatment and analysis—philosophical, poetical and scientific. In relativity it is a geometric dimension tied to space. From relativity we cannot get a direction for time—it is not sequential. When tied to space, as spacetime we would not know if time goes forward or backward because there is no forward or backward. It is only relative to something else. Only entropy, through the second law of thermodynamics gives time direction. It is remarkable that every child is convinced that time goes forward into the future, and ignores the fact that "forward" is relative to one's self.

The second law of thermodynamics is assumed to be an unchanging law of the universe. According to this theory we require a cycling universe. This means that entropy is reversed after the equilibrium between the kinetic system, (the system we observe and know) and the entropic system of spacetime. The latter is this theory and was invented to complement our standard model physics. Entropy, according to this theory is only applicable in an expanding universe.

Does time reverse in the contracting epoch? Does its direction change by π radians (180°)? The answer is no! The direction of time changes only $\pi/2$ radians (90°). FIG. 2 below shows the relationship of time and space in an expanding and contracting universe.

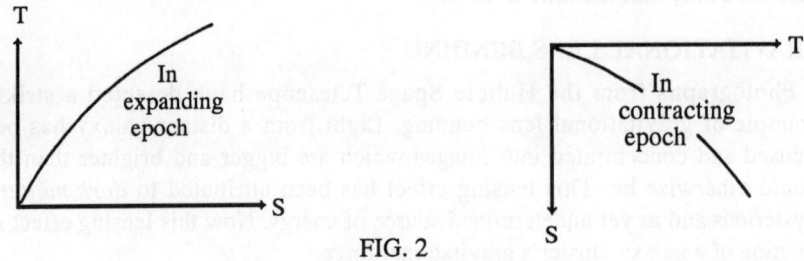

FIG. 2

This should be no more confusing than the direction of the magnetic field in relation to the electric field in Maxwell's electromagnetic waves. Magnetism is not opposite of the electric wave, but at a right angle to it.

$$\frac{K_E}{K_M} = \left[\frac{\ell^2}{t^2}\right] = c^2 \qquad \text{where } K_E = \frac{1}{4\pi\epsilon_o} \text{ and } K_M = \frac{\mu_o}{4\pi}$$

Fortunately our civilization is a factory producing vast amounts of observations. One of these observations was that in Pion-Proton scattering the N series resonance corresponds to 10^{-23} seconds. This is the shortest time ever deduced from observation. The shortest time interval deduced from this theory and applicable to the present time is:

$$t_{now} = 1.3512134 \times 10^{-43}$$

If at any future observation a time interval less than t_{now} is discovered this theory would be demolished.

WHAT HAPPENS TO TIME AT THE BIG BANG PART OF THE CYCLE?

At the end of the next contracting epoch, the orientation of time and space does change by π radians and a new expansion epoch begins. Suffice it to say that in the antimatter universe the direction of time and space will be meaningless to the life forms which could evolve then. The spacetime of Einstein still has no arrow of direction and only the entropy can give this clue. The next expansion epoch should simply repeat the entropy of the antimatter universe just as ours has given direction to time by virtue of the kinetic system pouring energy into the entropic spacetime.

BLACK HOLES?

Headlines from the world of astronomy: *"SPACE TELESCOPE CONFIRMS THEORY OF BLACK HOLES"* describes an invincible force gobbling up matter in space. A former skeptic declares: I do believe there is a black hole there."

Dr. Richard Harms is quoted: "If it is not a black hole it must be even harder to understand with our present theories of astrophysics."

We congratulate Dr. Harms for his restraint and we congratulate the Hubble Space Telescope for measuring the velocities of the spinning elliptical galaxy's (M87) gases.

The story goes on: "Having applied straightforward Newtonian physics the calculated mass at the center of the galaxy is of such magnitude that only a black hole would qualify." No congratulations there! Why?

As Vera Rubin wrote:"...perhaps as much as 99% of matter in the universe is not radiating at any wavelength. This dark matter...is detected by gravitational attraction on matter which we see." Vera Rubin also spoke of the inescapable conclusion that dark matter must exist.

According to the theory of entropic spacetime (EST), space is not a vacuum at all. Instead the theory states that space is tremendously dense. EST functions as dark matter. To calculate by Newtonian means that the mass of the central, light emitting portion of the galaxy must be wrong. Hence, the conclusion that a black-hole exists must be equally wrong!

EST is dense enough to guarantee a cycling universe. Dense enough to be the long sought DARK MATTER. Now with the "black hole" discovery this description of space can be applied to M87 as the reason for keeping the gases from flying away—without a black hole at the center.

Somehow, there has been a general belief that matter can be collapsed to greater densities than the nucleons themselves possess. This has led to the concept of singularity—a condition at the moment of the Big Bang.

The EST theory points otherwise; there is no singularity, there are no black-holes, and space itself functions as Dark Matter and as the unique basis of gravity.

CHAPTER VI

MODEL MAKING OF RADIATION AND MATTER IN A SHRINKING UNIVERSE

In the following pages I will try to illustrate and give specifications to a series of vector bosons and subparticles of matter which grow logically from the entropic space time theory.

The building blocks of these are space itself which according to this theory has characteristic length ℓ time t energy E and charge q.

Further, according to this theory, the conversion of space into radiation and matter can only occur in the cosmological epoch where the universe shrinks. Since we acknowledge our existence in an expanding epoch the models which are illustrated here cannot possibly be in existence in our present state. What then, you may say, is the purpose of this exercise?

The purpose of this is to give meaning to a new cosmology. This Cosmology consists of a cycling universe, which expands and then shrinks only about 37 billion times. This is far from the present concept of a "singularity" which exploded to create a universe. Our astronomers are having a difficult time scheduling a time where galaxies are being created in our expanding period.

In this new cosmology we find that space is not a vacuum but a rather dense body of energy and charge. Moreover, space is far more dense than any matter. This accounts for the gravitational effects of "dark matter" which is space itself. The big bang in this cosmology is a matter-antimatter explosion, not a mystical singularity. It also helps us explain that the galaxies which exist today were created in the pre-big-bang epoch during the shrinking period. This also helps us explain why we do not find any antimatter; it was burned away at the big bang. The galaxies and other matter which we find were the surplus matter and existed from the day of the big bang.

THIS IS THE JUSTIFICATION FOR THE MODEL MAKING IN A SHRINKING UNIVERSE.

Model making can best be done if we have material which lends itself to analogical treatment. In search for an elasticum, the rubber band seemed to suit the purpose. We will start with a flat untwisted rubber band. One side will be positively charged, the opposite side will be negatively charged. The thickness of the band will represent time, (or a timelike distance). The band itself will represent length ℓ. The internal stretch of the band, represents energy.

MODEL 1. THE SPECIFICATION FOR THE SMALLEST VECTOR BOSON

This is the first conversion of a scalar boson into a vector boson. In other words this is the lowest radiation photon possible:

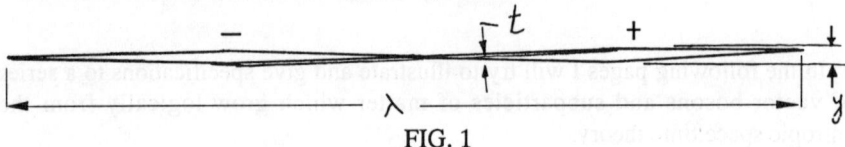

FIG. 1

Imagine taking a long, thin rubber band and giving it a single 2π twist. The action for this is h. A single twist is (1), this represents the frequency. A wavelength λ will immediately appear.

The band has a width, which we call y. In terms of a wave this is an amplitude.

All the energy of a photon is contained in the wavelength. Since the energy is $E = h\omega$ and $\omega = 1$ the energy must travel from one end of the wavelength to the other at velocity c. Therefore the longest λ is 2.9979246×10^8 m.

This is contrary to the present belief that a wavelength can be as long as the universe itself.

The basic relations of space are

$$E = \frac{q^2}{2\ell} \qquad \ell = \lambda y \qquad \lambda = \frac{hc}{E} \qquad y = \frac{q^2}{hc} = \frac{E\ell}{hc}$$

SPECIFICATIONS

$E = 6.6260755 \times 10^{-34}$ J

momentum $p = E/c$

$\ell = 1.2687795 \times 10^{-45}$ m

$y = 4.2321928 \times 10^{-54}$ (unitless)

$q = 1.2966903 \times 10^{-39}$ C

$t = 4.2321929 \times 10^{-54}$ s

MODEL 2. A BEAM OF PHOTONS IS SHOWN ON THE SKETCH BELOW

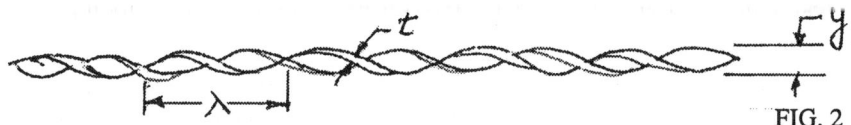

FIG. 2

A SPECIFICATION FOR A TYPICAL VISIBLE LIGHT PHOTON

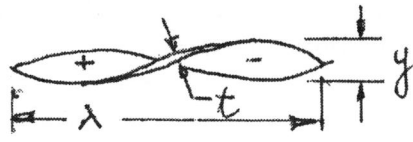

FIG. 3

$E = 1eV = 1.60217733 \times 10^{-19}$ J

$\omega = E/h = 2.4179883 \times 10^{14}$ Hz

$\lambda = c/\omega = 1.2398425 \times 10^{-6}$ m

$\ell = E\ell_0/E_0 = 3.067894 \times 10^{-31}$ m

$y = E\ell/hc = 2.4744224 \times 10^{-25}$ unitless

$t = 1.0233393 \times 10^{-39}$ s

$q = 3.135382 \times 10^{-25}$ C

Note: As energy increases in a vector boson, the wavelength λ decreases and y increases. $\ell = \lambda y$ and is proportional to E.

MODEL 3. SPECIFICATION FOR VECTOR BOSON CAPABLE OF PAIR PRODUCTION

The minimum energy requirement for pair production is 2x rest energy of electron. The rest energy of an electron is .51099906 MeV.

$E = 1.021998$ MeV $= 1.6374223 \times 10^{-13}$ J

$\lambda = 1.2131553 \times 10^{-12}$ m

$\omega = 2.4711797 \times 10^{20}$ Hz

$\ell = 3.1353822 \times 10^{-25}$ m

$y = 1.2922426 \times 10^{-13}$ unitless

$q = 3.2043546 \times 10^{-19}$ C

$t = 1.0458509 \times 10^{-33}$ s

Upon collision this vector boson splits in two, possibly in the manner shown on FIG. 4

The value of λ doubles to $2.4263106 \times 10^{-12}$ which happens to be the exact value of the Compton wavelength λ_c.

The frequency ω is half that given above. ℓ, being proportional to the energy is also half that given above. The y dimension is also reduced in half—which suggests that the split takes place longitudinally, as the diagram indicates. The charge is precisely that of the classical "elementary charge". This value is not recognized in quantum mechanics; only the specific charge of the electron is still recognized (the ratio of charge to mass).

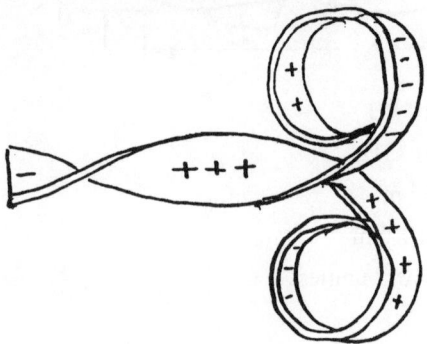

FIG. 4. A concept of how a vector boson splits in pair production

MODEL 4. AN ELECTRON

The conversion of a vector boson into an electron or positron can be demonstrated using a rubber band. First, color one side of the band to distinguish one side from another. Say the colored side is positive and the plain side is negative.

Twist the band 2π turns. Hold the band in such a way that only one wavelength is visible. Bring the ends together, thus simulating a collision. A loop, without twists will result. The concept here is that as a photon, the opposing charges are in a state of being transported (along with the energy) as a torsion wave at velocity c. This makes the charges appear neutral. When the loop is created it is no longer transported as a vector boson. The charge is now exposed and can be measured. Let's say that the "negative" side is exposed. Now we will show how a positron can be created.

Starting with the same band, held the same way as before, simulate the collision by bringing the ends together—except do it at an angle.

MODEL 5. PHASE CHANGE 1

In the previous models it has been demonstrated that as the energy level of a vector boson increases the wavelength λ becomes smaller and the amplitude y becomes greater. When the values of λ and y become equal a phase change takes place. The rubber band becomes tubular in shape. It is an interesting phenomenon; the phase change takes place suddenly, with a snap. The diameter of the tube is much slimmer than the y dimension.

There is another characteristic of interest after phase 1 change takes place. You may recall that when we were making an electron loop from a vector boson by bringing the end together—as if in collision—well it was very easy to reverse the process by pulling the loop apart and getting a vector boson again. Now that we reached a higher energy level—high enough to collapse the walls of the band into a tube, the same process which made the electron loop-vector boson conversion easy, will now not work. Instead the collapsed wall remains collapsed whether it is pulled in or out. It is reminiscent of a particle which reacts very poorly with other particles—the neutrino. That is, of course, if you are inclined to use a rubber band as a guide to the secrets of the universe.

SPECIFICATION FOR A PHASE 1 VECTOR BOSON

$E = 1.7111873$ MeV $= 2.741625 \times 10^{-13}$ J

$\lambda = y = 7.2455101 \times 10^{-13}$ m

$\omega = 4.1376308 \times 10^{20}$ Hz

$\ell = 5.2497416 \times 10^{-25}$ m

$q = 3.7937874 \times 10^{-19}$ C

$t = 1.7511253 \times 10^{-33}$ s

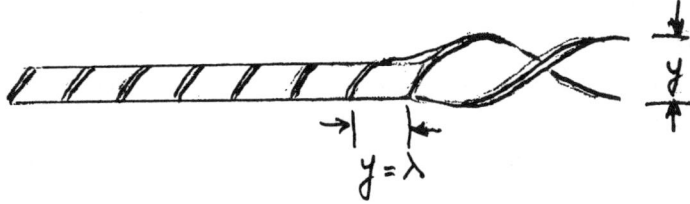

FIG. 5. A concept of a vector boson in the Phase 1 stage

60

PHASE 2. A WRITHE

Imagine what will happen to a vector boson in the tubular stage if we keep adding energy to it. We're bound to get a writhe as illustrated below:

FIG. 6. A concept of a vector boson in the Phase 2 stage—a writhe

The frequency of phase 1, ω, represents the number of turns in the photon. To get a writhe[1] from the tubular configuration requires additional energy; actually one additional turn (2π). How much energy is that?

$$\omega_1\hbar = 4.1376308 \times 10^{20}\ 1.05457266 \times 10^{-34} = 4.3634321 \times 10^{-14}\ J = .2723439\ MeV$$

or more simply $E_{\phi 1}/2\pi = 1.7111873/2\pi = .2723439\ Mev$

Thus the total energy in a phase 2 writhe is

$$E_{\phi 1} + E_{\phi 1}/2\pi = 1.9835313\ MeV = E_{\phi 2}$$

SPECIFICATION FOR A PHASE 2 BOSON

E = 1.9835313 MeV

λ = 6.2506828x10^{-13} m

ω = 4.7961554x10^{20} Hz

ℓ = 6.0852641x10^{-25} m

q = 4.3975879x10^{-19} C

t = 2.0298256x10^{-33} s

As we shall proceed further in the model making we'll find that a phase 2 writhe is likely to be the equivalent of a gluon.

For the next step we shall need a whole beam of phase 2 writhes which may look like this:

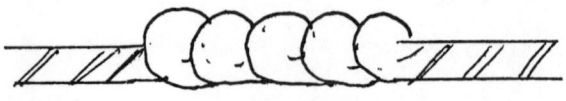

FIG. 7

PHASE CHANGE 3. A WRITHE OF WRITHES

Additional energy which acts on a photon beam made up of phase 2 writhes will convert it to a phase 3 writhe—essentially a writhe of writhes. If we follow the belief that according to the matriushka principle, it is likely that the large scale model will act like the invisible model, we find that about 6 phase 2 spheres will writhe into a phase 3 writhe. At least that is the visual indication. It should therefore contain a minimum of about 11.9 MeV. and it should also resemble a sphere of sorts.

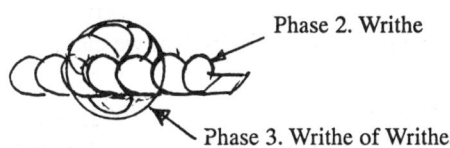

Phase 2. Writhe

Phase 3. Writhe of Writhes

FIG. 7. A concept of vector boson in the phase 3 stage, a writhe of writhes.

It must be getting obvious that we're now aiming at the construct of a quark from a number of phase 3 bosons, each of which is estimated to have at least 11.9 Mev. At present the mass of a quark (up and down) is given as about 310 MeV. This calls for an arrangement of 26 phase 3 bosons to make up a quark.

A CONCEPTUAL QUARK

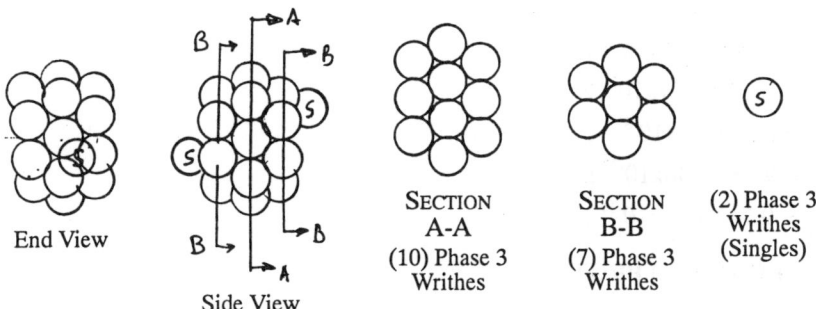

End View

Side View

SECTION A-A (10) Phase 3 Writhes

SECTION B-B (7) Phase 3 Writhes

(2) Phase 3 Writhes (Singles)

FIG. 8. A CONCEPTUAL QUARK

The 26 spheres arrange themselves easily to make up a quark. As the sketch above indicates; the order is 1-7-10-7-1, and the total is compact and simple.

I must admit that I was astonished by the ease with which these spheres fitted the energy requirements. I was even more astonished when the energies of the sum of 3 gluons and 3 quarks added up to be just about right to make up a proton. It seems almost as if the values are trying to encourage the concept.

SPECIFICATION – QUARK

Please note that I am not differentiating at this time between the UP and DOWN quarks. It is also uncertain whether the phase 3 writhe is exactly 6 phase 2 or slightly greater. The energy content of a phase 2 writhe is fairly solid -1.9835313 MeV.

6x1.9835313 = 11.901188 MeV

26 x 11.901188 - 309.43088, slightly below what is needed to make a proton. What we need is 310.77389 Mev; if we allow that there is slightly more than 6 phase 2 writhes in the making up of a phase 3 writhe—so if we assume that the number is not an even 6 but 6.0260416 than we can have an exact energy count for a proton.

3 quarks @ 310.77389 MeV =	932.32167
3 gluons (phase 2) @ 1.9835313 MeV =	5.9505939
	938.272(26)MeV

In this case I cannot say that the adjustment is a way to "back in" to the right answer. It was obvious that writhe of writhes is not likely to be an exact number. The visual clue was six but there was no way to determine the exact number.

We can now write a specification for a quark:

$E_Q = 26xE_3 = 310.77389$ MeV

$\lambda_Q = 3.9895321x10^{-15}$ m

$\omega_Q = 7.5144767x10^{22}$ Hz

$y_Q = 2.389808x10^{-8}$

$\ell_Q = 9.5342146x10^{-23}$ m

$q_Q = 9.7439494x10^{-17}$ c

NEUTRONS AND PROTONS

Now that we have a quark let us develop a concept for a neutron. The idea of up and down quarks is left out of this; the reason is that fractional charges of $-1/3$ and $+2/3$ associated with the present thinking has an alternate to what appears to be a simpler accounting for the charged state of protons and neutral state of the neutrons. In this theory, the quarks for neutrons and protons are the same. We can blame the model making from rubber bands for this—there is no obvious partial charge on the quark.

What this model making exercise must account for is the difference in the rest energy content of the neutron and proton and how can the exact countercharge of the electron end upon in the proton, and be missing in the neutron.

Neutron rest energy	939.56563 MeV
Protron rest energy	938.27231 MeV
Difference	1.29332 MeV

It should be noted that the energy difference is less than required for a phase change but ample for a pair-production photon.

Let us suppose that a neutron is made up of (3) quarks and (3) gluons (phase 2). However the way they hang together is not the most compact arrangement. Compactness can be attributed to the location of the single phase 3 writhes. (Each quark has 2 of them 1-7-10-7-1) We are talking about the singles which could be attached in a number of places. The gluons are located between the singles as shown in the sketch on next page.

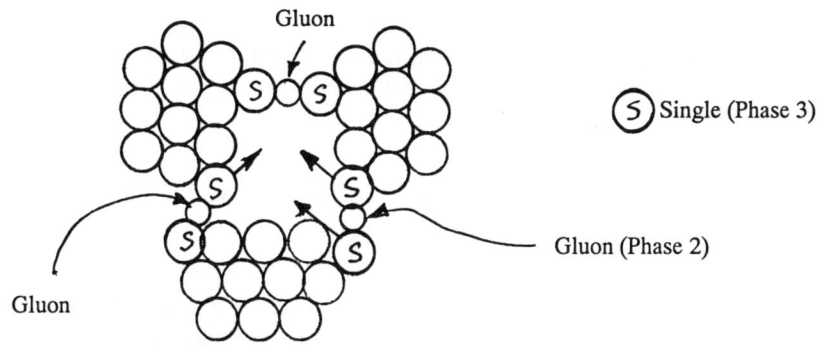

FIG. 9. A CONCEPTUAL NEUTRON—NOT THE MOST COMPACT
ARRANGEMENT

It is well known that a single neutron will decay in a relatively short time into a more compact proton.

This is generally written as $n \rightarrow p + {}^- e + \bar{\nu}$.

But we could write this as a pair production relationship, where the positron never leaves the nucleon and becomes a proton.

$n \rightarrow (p^+ e) + {}^- e + \bar{\nu}$

The energy distribution would leave .27132 MeV for the neutrino to carry off. The satisfying aspect of this is that it accounts for the exact positive charge of the proton—as far as I know no other theory is able to do this.

64

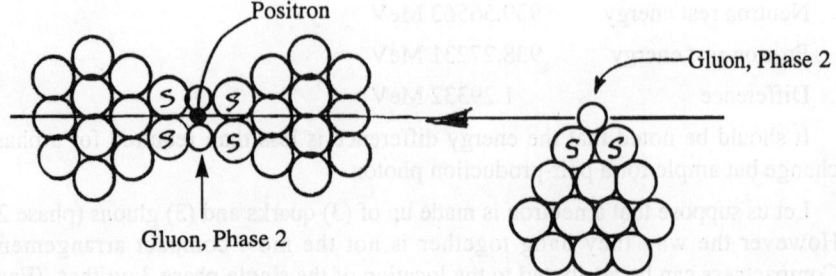

Positron

Gluon, Phase 2

Gluon, Phase 2

This Quark is removed for clarity
(belongs on top of cluster)

FIG. 10. CONCEPTUAL PROTON WITH POSITRON IN CENTER— THE MOST COMPACT ARRANGEMENT.

BOSONS AND FERMIONS

I liked to think that all things were either bosons or fermions. Bosons were photons, radiation, and were transported at velocity c, while fermions were matter, could never move at velocity c—but could only approach this velocity as its mass approached infinity. It made a nice clean division. But then things started to fog up.

We defined fermions as particles with half-integral spin in units of $\hbar$ which obey Pauli's exclusion principle.

We defined bosons as having a spin of 1 and abiding by certain statistics.

Then J.A. Wheeler cried out: "What place is there in quantum geometrodynamics for the neutrino? The only entity of half integral spin that is a pure field in its own right in the sense that it has zero rest mass and moves at the speed of light. No clear or satisfactory answer is known to this question today." Today the zero mass has been questioned by a number of physicists; consequently the velocity of c may also be questioned.

Then we have to consider the W boson-spin 1, of enormous mass, which decays into familiar leptons (the electron family) all of which are half spin and therefore fermions. Then we know, long ago, about bosons of gamma strength which could, upon collision, produce pairs of electrons. So we have conversion of a spin 1 particle into two 1/2 spin fermions. If this seems confusing, actually the overall complexity of the multitude of discovered particles hasn't even been touched. So where do we go from here?

As I see it now we have the following:

SCALAR BOSONS, spin 0 , which have a total of three dimensions, one of which is time. These bosons are space itself. Their length ℓ, as a space particle has a minimum and maximum dimension. So does its time interval. The specifications are as follows:

	Minimum	Present	Maximum ESTIM.
Length ℓ	$1.2687795 \times 10^{-45}$ m	$4.0508459 \times 10^{-35}$ m	4.717×10^{-35} m
Energy E	$4.1356692 \times 10^{-21}$ MeV	$1.3203963 \times 10^{-10}$ MeV	1.537×10^{-10} MeV
Temperat.	4.799216×10^{-11} °K	1.5347 °K	1.785 °K

The minimum temperature given above is the theoretical low limit, rather than the absolute zero often discussed. Any experiment which can bring the temperature below this value will disprove this theory.

The present temperature given above is to be compared with the 2.7 °K temperature measured of the background radiation. Had it worked out to be greater than 2.7 K this theory would be wrong. We are comparing the temperature of vector bosons or radiation spin 1, with scalar bosons—which is the temperature of space itself—spin 0.

From the description of space above, we see space as an accumulation of potential energy in a cosmology which is currently expanding. Notwithstanding its very low temperature, it contains a vast amount of energy. When the background radiation temperature is equal to the temperature of space itself, the entropic process must necessarily cease. Space will then contract and this energy will be released.

The building blocks of the universe will then go to work, as the model making section has indicated. All the particles which have been observed—the short lived and the stable ones will then be created.

We have on the stable list:

Scalar bosons, spin 0 (space)

Vector bosons spin 1 (radiation)

Neutrinos ν_e from electron reactions spin 1/2

Neutrinos ν_μ from muon reactions spin 1/2

Electrons ^-e and ^+e spin 1/2

Neutrons and Protons

All short lived particles will decay in complicated decay patterns into these so-called stable types.

What has the model making process contributed?

We have observed, through model making, that there are other states which have never been pointed out. The phase changes occur as the energy content of a vector boson increases.

The phase 1 change occurs at 1.71 MeV when a tubular shape occurs.

The phase 2 change occurs at 1.98 MeV when the tubular boson writhes.

The phase 3 change occurs when a series of phase 2 writhes make up a writhe of writhes. We found that it takes 6.026 phase 2 writhes to make a phase 3 writhe. The energy count then is 11.95 MeV for each phase 3 writhe.

Now we have the brick and mortar to build all sorts of particles. We found that a phase 2 writhe fitted the gluon. 26 phase 3 writhes made up a quark. 3 quarks and 3 gluons made up the neutron-proton particle, just as predicted by Gell-Mann and company. (Only by a completely different route).

When does a spin 1 boson become a 1/2 spin particle?

In the model making we show that simple vector bosons (by simple we mean one which has not undergone a phase change) can make electrons. This process requires a collision and a looping.

Further up the ladder of energy content, phase changes occur. No collision is required. The energy itself forces phase changes. I would identify the force of this energy as the STRONG FORCE. In the case of the electron there is no strong force because the collision itself relaxes the tension within the boson. What is new here is that we are saying that the strong force exists in both mass particles as we expect but that it also exists in the writhes of the bosons. When they loop as in the creation of a neutron-proton then the particle is definitely a 1/2 spin item.

It becomes apparent that strings or beams of radiation are spin 1. When they loop, they become spin 1/2. This includes electrons and long chains of writhes (as quarks) when they loop, they become neutron/protons.

Now a loop is not a writhe. A loop is closed and a writhe is still open. Loops are 1/2 spin. Bosons, as open strings or beams are spin 1.

THE NEUTRINO

In the model making description given above, starting from the lowest energy vector boson all the way to the quark and higher, somewhere along the train of models lies something which will answer the requirements of the neutrino.

What are the qualities which a neutrino must fit?

It must be "massless" or nearly massless, because some electron do emerge with the maximum energy allowed in the β decay process.

Bosons are considered massless, not because they do not hold energy but because they have no "rest" energy. We merely attribute a momentum to them. But neutrinos cannot possibly be bosons, because bosons have spin 1, where neutrinos are given spin 1/2.

According to the conclusion reached concerning loops, to have 1/2 spin it must be a loop. A neutrino, in the world of rubber band models, must be part boson and part fermion; sort of a half breed which is composed of several wavelengths of radiation, in the center of which is a loop. Maybe the loop is not firmly closed, but looped enough to qualify for some small amount of rest energy.

A clue from the model is that when a neutrino model is pulled apart, the center loop does not relinquish its shape. The same thing done to an electron loop immediately transforms it into a vector boson. This would account for the neutrino's reluctance to react with other particles.

The bosonic wavelength of the neutrino component would readily account for its ability to carry off varying amount of energy. There would be a minimum amount of energy which a neutrino must carry off, that is at least $2\hbar$ + mass of the loop. So in those β decays where we think that the electron has carried off all of the energy, it must be still something less than all. These cases are quite rare anyway but it would be interesting to see just what that would be.

It must have also become apparent that the neutrino loop can be right or left handed. Since Dr. Goldhaber's work in '58, we know that all neutrinos are left-handed and all antineutrinos are right handed. This concept of helicity fits the model making concept well.

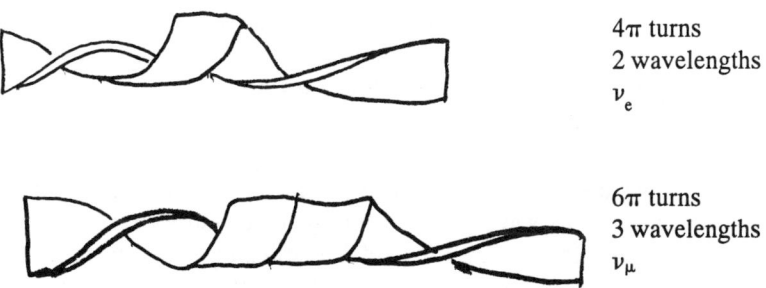

4π turns
2 wavelengths
ν_e

6π turns
3 wavelengths
ν_μ

FIG. 11. CONCEPTUAL MODELS OF ELECTRON AND MUON NEUTRINOS

MODEL MAKING AND QUANTUM MECHANICS

Wave and particle duality, the core of quantum theory seems totally remote from the model making of particles and radiation expounded in EST. In quantum theory the emphasis is in kinetics. For example the energy $mv^2/2$ of a free particle is defined so that it does not contain the rest energy.

In EST we work in elements which lead to the rest energy, with little interest in the kinetics. In EST the construct is from a flat potential field of charges to create particles of known rest energy.

DOUBLE SLIT EXPERIMENT—WAVE PARTICLE DUALITY

This experiment is one of the cornerstones of quantum theory. The elements of the experiment are: a source of light or of electrons, a screen with a double slit and another screen provided with a detecting device to evaluate the nature of what passed through one or both slits. No attention is paid to the nature of space through which the light or the electrons have passed. The gas particles of air are ignored as well.

The detecting screen shows a diffraction pattern (rings) when both slits are open. When one slit is closed merely a distribution is shown. The ultimate test is when only one particle, say an electron, is passed with both slits open. (You can imagine the immaculate care required for this test.)

Within the rigors of quantum theory the traversing particle is considered a wave, but when it is deposited on the screen it does so as a particle. This includes the single electron which is considered to have passed through both slits. The classical viewpoint, on the other hand would lead us to believe that each particle is an entity and must pass through one slit or the other. EST views both radiation and particles not as entities, but torsion waves in the space called EST. In model making a flat (untwisted) space (actually surfaces of charges) is twisted by the release of energy-charge when the universe contracts. The twisting is in terms of h—each twist accounting for one $\hbar$, the angular momentum of h. An electron attains the rest energy required after 1.23×10^{20} twists which we now call its frequency. It is still a vector boson (photon or γ ray) until a looping occurs. This happens when the photon collides with another particle.

As a vector boson the energy is transported at velocity c. Only local vibrations of the charges are required to create a boson and propel it at velocity c. The collision which brings about the looping (and the creation of an electron) does not stop the movement of its wave. Now it spins @ αc, a velocity 1/137c. The spinning of the charges create a magnetic field. The spinning of the electron does not tear it loose from the surrounding space quanta—rather, the EST bound to the electron is restored to the original state every 4π radians of spin. (P.A.M. Dirac showed how this is done.)

Back to the double slit experiment! In the standard model of physics the photon or electrons travel in air, a gas inhabiting a vacuum, or in a void, depending on the experiment. EST is a new element to consider. Wavelengths, like a party ribbon appears to move when it is only twisting. The surrounding EST quanta are affected by the spinning vector. Again we have a disturbance restored every 4π turns.

According to the model making each photon is one wavelength. When that photon has made one turn it has also moved one wavelength away. That is only 2π radians; so it is every 2 wavelengths before 4π turns are made. This disturbance brings about a transverse wave whose wavelength should be twice that of the deBroglie wavelength—in photons, electrons or other particles. The photon is a wave, but the wavelength which contains all of the photon energy is also a particle—so there we have the duality. The electron is similar to the photon except it is a loop in the EST. In each case they can create transverse waves in the EST. (Real waves). This spares us the agony of saying that the probability waves (so named by Born) actually cause diffraction or interference.

This spares us the need for saying that a single electron can pass through both slits. The electron can be a particle while traveling through a slit while the transverse waves in the EST produce the effects. This seems hardly to interfere with Schrödinger's ψ wave function to determine probabilities. It does not interfere with Heisenberg's uncertainty principle which states that location and momentum cannot be determined simultaneously. Quantum mechanics thrives in the kinetic system. EST is the theory of space. In quantum mechanics, when one speaks of the states of an electron, one means the kinetic energy associated with that electron. In EST we are only concerned with the rest energy of a particle and how it is connected with the space from which it is made. The rest energy of a particle does not vary, not even at relativistic velocities. The mass of the particle may increase since it is the sum of the rest energy and the kinetic energy.

There are some differences in the notations:

QUANTUM	EST
$E = h\nu$	$E = h\omega$ ω is potential frequency or
$E = h\omega$	angular frequency when converted
$\omega = 2\pi\nu$ angular frequency	There is no ν
$p = h/\lambda$	$p = E/c$ momentum as it applies to vector bosons. $E_o = h$ $\lambda_o = c$
not used	EST is $E/\mu_o \epsilon_o c^2 = E$
$k = 2\pi/\lambda$ wave number	not used
$k = p/\hbar$ wave vector	not used
$\hbar = h/2\pi$	angular momentum was shifted to $\ell_{now} = \ell_p (2\pi)^{1/2}$ (ℓ_p is Planck length)
not used	$\ell = \lambda y$ where $y = q^2/2hc = E\ell/hc$
$\lambda = hp$	$\lambda = hc/E$
ψ wave function	not used, EST is ubiquitous, location, momentum and probabilities are of little interest
not used	$E = q^2/2\ell$ the relation of energy, charge, length.

WHY A PHOTON KNOWS NO TIME?

In the model making we show a photon as a twisted plane of charges. The dipoles are separated by a "time" dimension.

In the scalar state of EST, where we have untwisted planes of charges, it is obvious that when a body transgresses the planes, it requires time to do it. The time has no direction or flow. We, on the moving body accumulate time crossings. So does a body which comes at us or a ray of light which comes at us—when we do the observing.

The photon itself knows no time because the time dimension which is twisted is the same time. This photon model enables us to see why a photon knows no time.

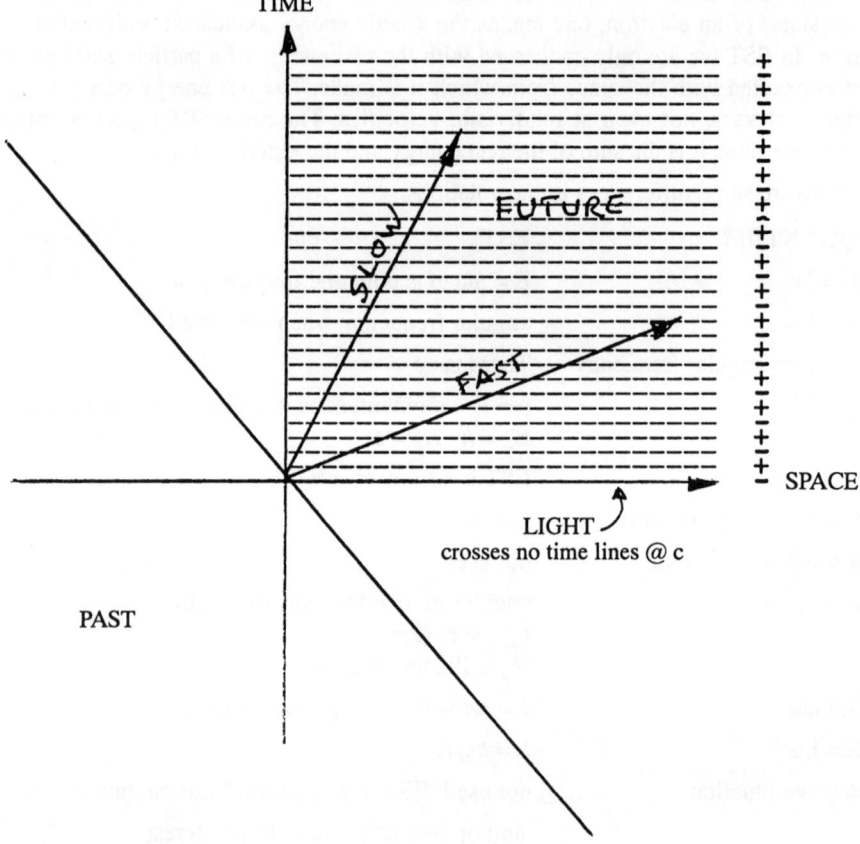

FIG. 12. A light cone arranged with EST time lines

TIME AND VELOCITY IN A DISCRETE UNIVERSE

Starting with the premise that "a photon knows no time" really means that flight of a photon through EST, crosses no time lines.

In model-making we show that a photon carries its own time.

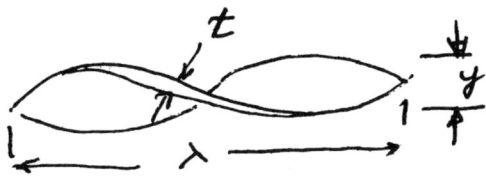

FIG. 13.

From this, as a beginning we must conclude that the more time lines are crossed, the slower the body's (or the inertial frame's) movement.

Using the values from the EST TABLES we have:

ℓ_{now} = 4.0508x10^{-35} m as the current distance between space quanta in the spacelike direction and

t_{now} = 1.3512x10^{-43} s as the current time interval or the dimension in the time-like direction.

When we say that the maximum velocity is c, we are really saying

v_{max} = c/ℓ_{now} = 7.400x10^{42} spacelines / time line.

No time lines were crossed.

The converse—when we are considering the minimum velocity:

v_{min} = c/t_{now} = 2.219x10^{51} timelines / space line.

No space lines are crossed.

The time interval t and the space interval ℓ are dimensions of the discrete universe and are variable. For example at the time of the big bang the maximum velocity was c/ℓ_o and the minimum velocity was c/t_o.

We can see from the above that:

v_{max} = 1/t   (ℓ = ct)

v_{min} = c^2/ℓ (ℓ/c = t)

ℓ and t are the vintage length and time for the applicable period.

The minimum velocity is rather fascinating because we don't hardly think about it. The minimum velocity, in this scheme of things means that something is not subjected to any movement in an absolute sense. The actual movement is limited to the size of ℓ and this is tiny indeed. Even more startling is that the units for minimum velocity are not ms^{-1}, but ms^{-2}. The latter is the signature for acceleration, not velocity.

Suddenly this makes sense: An object which is "still" in an absolute sense can only be subjected to acceleration.

CHAPTER VII

SURFACES

In the modeling of radiation and mass particles one could not miss the meaning of ℓ, a length which was measured in meters along λ, the wavelength and y, a unitless amplitude. This forces the dimension of ℓ to be in meters even though it looks like a surface. We called ℓ gravitational length to distinguish it from the ordinary meaning of length. ℓ has peculiar characteristics. It increases in proportion to the energy it contains, but the λ, the potential wavelength gets smaller while the amplitude y gets larger.

$$\frac{m_P}{m_e} = \left[\frac{\ell_P}{\ell_e}\right]^{1/2} = \frac{\lambda_e}{\lambda_P} = \left[\frac{y_P}{y_e}\right]^{1/2} = \frac{t_P}{t_e} = 1836.1527$$

When compared to the mass ratio of the proton to the electron the length and the amplitude change as the square but the wavelength is inversely proportional. The Compton wavelength is obtained by dividing ℓ/y.

The scalar boson aspires to be a building block for all things. First by way of twisting, (angular momentum) then through phase changes. Vector bosons are created first and after that, through the phase changes into fermions. This necessitates going more deeply into the meaning of the word surfaces as compared to the meaning of length.

With the expansion of space we have dilation of both length and time; we also have simultaneous growth of energy content and charge. The length concept of space and subsequently associating this with a surface is the part we will deal with here.

To begin with we said that a constant k_{dil} exists, which permits finding the length if the energy is known.

$$k_{dil} = \frac{E}{\ell_{dil}} \quad \text{where } \ell_{dil} \text{ is the dilated length}$$

We have to call on classical values to support some of these observations not just to duplicate the values but in addition gain further insights.

This theory postulates that there is a surface associated with every particle or combination of particles. For example there is a surface for an electron and a surface for a proton. How do these surfaces act in the Bohr radius a_o?

But first, how shall we transform the length ℓ into a spherical surface with a radius r_ℓ?

$$r_{\ell_e} = \left[\frac{\ell_e}{4\pi}\right]^{1/2}$$

Our topology holds that the charges are on the ℓ (as a scalar boson) and then becomes twisted up (angular momentum) as a vector boson, but the charges remain in place. When an electron is formed, a loop is created on an untwisted surface. The untwisting is the result of a collision of the vector boson.

In the proton the length has undergone three phase changes and is a cluster of writhes, in the center of which is a positron. All the charge dipoles are obscured so only the positron prevails. The positron exists by virtue of the decay of a neutron which resulted in pair production in the decay process, where an electron and a neutrino were created and released, while the positron remained to give the proton the discernible charge.

In Fig. 1. is shown a proton radius and an electron radius which are spaced from each other at the Bohr radius a_o. It is not important here to ask why the quantum mechanics have all but abolished these values. We are not interested here in the kinetics of the electron and proton. We are trying to make a continuity between a scalar system of space by bridging into the underlying logic of the statics which could then be combined with quantum theory.

The distance between the surfaces of the proton and the electron is ce. The product of c and the charge of the electron. The units are coulomb meters/second.

$$ce = a_o - \left[\left(\frac{\ell_{pr}}{4\pi}\right)^{1/2} + \left(\frac{\ell_e}{4\pi}\right)^{1/2}\right] = 4.802 \times 10^{-11}$$

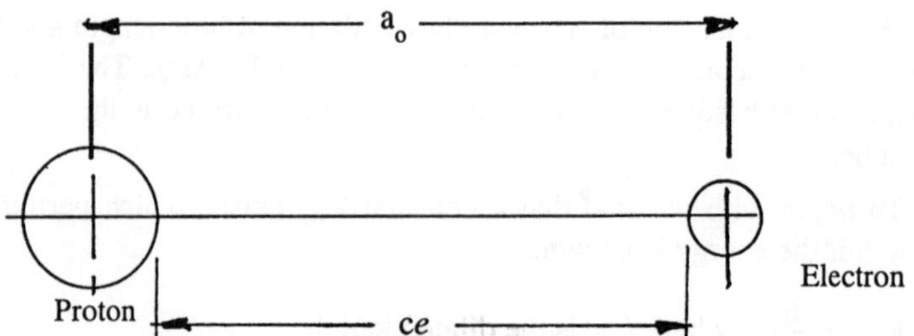

FIG. 1

Such coincidences don't occur randomly! Furthermore, if we calculated the attractive force of the proton and the electron over a distance (ce)

$$F = \frac{K_E\, e^2}{(ce)^2} = \frac{\mu_o}{4\pi} = k_M = 1.000 \times 10^{-7} \text{ N}$$

The simplicity of this relationship adds believability to the idea of surfaces. μ_o, the permeability of the vacuum, should be called the permeability of EST. This leads to the belief that the surface of a particle is the location of the charges. This is applicable whether the charges are discernible like an electron or proton or appear to be neutral, like the neutron. The fundamental charges are still there—the discernibility is only a matter of the topology of the particle.

One might say that the equation of F (above) is a tautology, because e^2 cancel out and $K_E/c^2 = K_M$. But the ce equation cannot be a tautology.

THE COULOMB LAW BREAKDOWN

In classical physics it has often been said that the Coulomb laws breakdown in the close vicinity of charged particles. What new light can the entropic spacetime theory shed on this situation? We have shown that, in accordance with this theory, the charges are located on the ℓ_{dil} (dilated length) of the space quanta.

The charge of an electron should therefore be located on ℓ_e.

$$\ell_e = \frac{E_e}{K_{dil}} = 1.5676911 \times 10^{-25} \text{ m}$$

The radius of a sphere whose surface is ℓ_e:

$$r\ell_e = \left[\frac{E_e}{4\pi}\right]^{1/2} = 1.1169283 \times 10^{-13} \text{ m}$$

The so-called classical radius of the electron $r_e = 2.8179409 \times 10^{-15}$ is considerably smaller than $r\ell_e$. To pursue this idea further we require a number of assumptions.

We are saying that:

- All charges are indestructible; the law of preservation of charge is applicable to dipoles and individual positive and negative charges.

- All charges of looped vector bosons (mass particles) reside on a spherical surface of radius $r\ell_e$.

- All particles, (i.e. mass particles or bosons) have a charge surface. (Not necessarily discernible.)

- The surface density of the charges is the same on all particles and on space itself. The charge density for all elementary particles is 1.0219981×10^6 C m^{-1}. This includes all neutral particles and those of discernible positive or negative charges. This charge density was derived from:

$$\frac{(2E\ell)^{1/2}}{\ell} = \frac{q_{now}}{\ell_{now}} = 1.0219981 \times 10^6 \, Cm^{-1}$$

The fact that the charge surface is larger than the supposed mass radius r_e is not in conflict with present thinking. This is because modern physics has rejected the dimensional aspect of the electron radius. The only allowable strictly defined number in the electron in modern physics is the specific charge of the electron—the ratio of charge to mass.

It all boils down to this: The classical observations have been that the Coulomb laws break down in the vicinity of the charged particle mass. This mass is described by the radius r_e. Since the modern physicists don't acknowledge this radius as an actual dimension then they have no problem with the Coulomb law breakdown.

Here we are stating that we don't know what r_e is. We give however another dimension for the radius of the charge sphere about an electron and say that the Coulomb laws will terminate at that radius. Moreover this radius is larger than the classical r_e. We agree with the classical supposition that Coulomb laws break down in the vicinity of charged particles and we give the exact dimension of where Coulomb laws should stop. From older literature the Coulomb breakdown occurs in the 10^{-14} region which is close to our dimension.

This theory states that Coulomb laws should cease for "charged" particles (by the conventional definition) at:

$$r_{\ell_e} = 1.1169283 \times 10^{-13} \, m \text{ for the electron and}$$

$$r_{\ell_p} = 4.7860772 \times 10^{-12} \, m \text{ for the proton.}$$

MORE ON SURFACES

Recall that we call the two dimensional ℓ a length most of the time. It is two dimensional because $\ell = \lambda y$ but y is a unitless amplitude while λ is an ordinary wavelength. In the Coulomb Law Breakdown we treat ℓ as a surface so that we can get a charge radius. This appears to be a dilemma and requires further thought.

A Euclidean line has no thickness—it has one dimension. Our length has two dimensions.

The conventional vector boson (photon) is given a wavelength and frequency but no amplitude. This is a bit of a mystery because we also think of most sine curves as having a wavelength and an amplitude. In our current convention we don't offer pictures of a photon in any of our physics books. But here we have pictures of radiation which include both dimensions (λ and y) and time as well. The reason for this exercise is to show that we can treat ℓ as a surface and derive interesting results.

We described EST as three dimensional, one of the dimensions is time. This means that the other two dimensions are incorporated in ℓ, the gravitational length.

There is another way of looking at this problem. At present the general belief is that the universe is a continuum of space. We have energy as a discrete exception. Energy comes in bundles but space is not. In effect, since this theory has placed energy as the real meaning of space, therefore everything is quantized, including time, which is ℓ/c.

Now, duality is nothing new. Quantum theorists have made particles and waves a duality. This is a Zugszwang—a forced move—in order to make sense of the whole. We also have a forced move: the need to define ℓ as a two dimensional beast (except, as you might have noticed in the attempts to volumize space).

THE CLASSICAL RADIUS OF THE ELECTRON:

The rejected r_e still has a direct connection to surfaces in this theory: $r_e = 2k_E\ell_e$. Moreover it is connected to the ratio of the electron charge e to the surface density of charge ρ_q.

$$r_e = \frac{2k_E e}{\rho_q} = 2k_E\ell_e$$

If we wanted the internal electrical force of the electron using our topological model then:

$$F_E = \frac{k_E e^2}{\ell_e} = 1.4716416 \times 10^{-3} \text{ N}$$

We can substitute $2E$ for e^2/ℓ_e

$$F_E = 2k_E E_e \text{ compared to}$$

$$r_e = 2k_E\ell_e$$

We can say that: The electrical force is to the rest energy of the electron what r_e is to the electron length (charge surface). We could also say:

$$\frac{F_E}{r_e} = k_{dil} = \frac{E_e}{\ell_e}$$

The length of the electron can also be described in terms of the Bohr radius a_o and the fine structure constant α.

$$\ell_e = \frac{a_o\alpha^2}{2k_E}$$

It just so happens that $r_e = a_o\alpha^2$.

UPON THE ANOMALY OF THE ELECTRON MAGNETIC MOMENT

Can we obtain through this theory results which QED holds among its crown jewels?

The answer is yes and no. Yes because we clearly get the right answer, but no because we fail in the accuracy of the QED method. The limits of accuracy are imposed by our failure to get the value of Newtonian G close enough. The uncertainty there is something like 128 parts per million. Our value of ℓ is dependent on Planck's ℓ and it in turn is dependent on the accuracy of G.

When Bohr created the expression for the electron magnetron

$$\mu_B = \frac{e\hbar}{2m_e} = 9.2740154 \times 10^{-24} \text{ JT}^{-1}$$

It became the basis for the magnetic moment at that time. Later this value was found slightly incorrect and the empirical electron magnetic moment μ_e, was calculated to be $9.2847701 \times 10^{-24}$

$$\frac{\mu_e}{\mu_B} = 1.001159652193 \text{ is the ratio, as calculated at present.}$$

This is the basis of the electron magnetic moment anomaly. μ_e is considered an empirical constant and has no equation tying it to known basic constants, but it was calculated with great accuracy to experimental data through QED.

THE ANOMALY, AS SEEN THROUGH THE ENTROPIC SPACETIME THEORY

The entropic space theory allows non-spin bosons at many energy levels, as long as ω is an integer. In a boson with an energy equal to E_e the electron, the value of ω_e is:

$$\omega_e = 1.2355898 \times 10^{-20}$$

where ω_e represents the number of times the fundamental energy E_o is in E_e (also called the potential frequency of a scalar boson).

The electron charge, in Coulombs is $e = 1.6021773 \times 10^{-19}$ C, while in the equivalent boson the sum of the charges have exactly the same coulomb strength. The product of the coulomb charge times twice the number of charges (dipoles) can be expressed as $2e\omega_e$.

In the model making of the electron we see an untwisted loop with a length of ℓ_e, a width of y, and a wavelength λ_e exactly of the value of the Compton wavelength. ℓ_e is the product of λ_e and y_e. This is the non-kinetic (static) picture of an electron. The kinetic picture as viewed in the quantum mechanical scene is quite different. We know the electron spins and jiggles; the temperature of the electron is ~6×10^9 °K from just its rest energy. We may imagine that the loop is more like a sphere under those conditions. Spinning charges also have a magnetic effect.

The charges distributed on a spherical surface (whose perimeter is ℓ_e) would be on a charge sphere of radius:

$$r_{\ell_e} = \left[\frac{\ell_e}{4\pi}\right]^{1/2} = 1.1169(283) \times 10^{-13} \text{ m} \quad \text{where } \ell_e = 1.5676911 \times 10^{-25}$$

r_e which was once called the electron radius has a way of popping up every now and then as we mentioned before. EST cannot reject this value.

$$\frac{r_{\ell_e}}{r_e 2e\omega_e} = \frac{\mu_e}{\mu_B} = 1.0011(025) \qquad r_e = 2k_E \ell_e = 2.817941 \times 10^{-15} \text{ m}$$

What is lost in accuracy is gained in the fact that we can express this as an equation in known basics of this theory.

CHAPTER VIII

EST AND THE QUANTUM FIELD THEORY

At first it seems that the entropic space is worlds apart from Quantum Field theory. It becomes less so when we realize that it is only the approach which differs. In the Quantum Field theory all entities are described as fields; the photon is described as an electromagnetic field, the electron and proton are the electron field and proton field respectively. From the present quantum mechanical interpretation we must conclude the following:

- An electron which is isolated from interactions the uncertainty of its position is infinite. Its wavefunction ψ is a sine wave of wavelength λ extending throughout space. It would be in no case a localized particle.

- If an electron has disturbed an atom with Δx as dimension of the atom, there will be an uncertainty Δp in the electron momentum so it will spread the wavelength of the wavefunction to $\Delta\lambda = h/\Delta p$. This spread of wavelengths causes the formation of a localized wavepacket, reflecting a rough localization.

- When an electron is more specifically localized, like in a high energy collision with another particle then the wave packet becomes very localized and could be regarded as a particle—but its momentum would be very uncertain.

This picture of the electron wave makes nonsense of the simple Bohr picture of the atom. Until some act occurs which localizes the electron, there is no meaning to ascribe more detail to it. This is why in modern physics we have no attempt to give the electron any physical dimensions.

In the entropic spacetime theory, the fermions are seen as particles, but create waves much as the quantum field. This field is possible because of the nature of the entropic space. That is the part which differs between the theories. In Quantum mechanics, space is a vacuum, an unquantized continuum of nothing, which through necessity is given peculiar properties.

The unlocalized electron exists in the EST as a potential electron in a shrinking space, when entropy is reversed. The scalar bosons convert easily into vector bosons under those conditions. When a vector boson of suitable energy content collides electrons are created. The modeling indicated that it is looping of a straight wave that signifies the conversion to a fermion. In the present condition of the universe—that is in the expanding state—the energy level of scalar bosons is far too low to create any fermions. The energy of a current scalar boson is 2.11×10^{-23} J. To qualify for the energy level of an electron the scalar boson should be on the order of 8.18×10^{-14} J.

EST however will transmit any photon. Where would such high energy photons come from? They would necessarily come from existing particles interacting with each other (or isotopes breaking down).

Quantum mechanics has created a world of kinetics and a complicated system for dealing with kinetics. EST is strictly a theory of space; it touches on kinetics only when some interface is found. Considering that quantum theory considers space a void, there are only few interfaces.

It is the field which has the ubiquitous presence in space. It is the mathematics of this field which, although devoid of any real definition is able to make predictions.

The entropic spacetime is a quantized field; but then we define what a field is. It is charge, energy and distance and time from which fields may be formed. Until now fields were not given such a description. The entropic spacetime, as a field is made up of particles which we call space quanta or scalar bosons which we have already described in many ways. Suddenly this makes the entropic space theory much more in line with the present thinking. In the 'standard model' all entities are described as fields. In the entropic space theory all entities are particles and they create the same fields by stresses impinged upon the particle arrangement and hence we defined the meaning of the field. Space as a field or space as particles which create fields is not too remote from one another.

Now there is one more fundamental difference. Having quantized the field as an arrangement of finite particles, these particles are capable of having dimensions. EST makes the entire universe finite, in the name of conservation of energy.

The duality of particle and wave which is prevalent in the 'standard model' physics is not really in such conflict with the entropic space theory. It is natural that in a theory which sees everything in terms of waves one has no choice but to unite the two and call them the same.

OUTSIDE OF THE STANDARD MODEL

Outside of the standard model we have the Higgs field concept which is nearer to the entropic space theory than might at first be thought. So is the Higgs boson, which does not spin is like the entropic boson. Both provide a building block of everything in the universe. Another theory which has similarities to the entropic space theory is the Supersymmetry concept which claims that each fermion has an equivalent boson and vice versa. Here a boson which has the same energy as an electron is called a selectron. In Supersymmetry a fermion may change into its equivalent boson and nothing observable will happen!

A LITTLE MORE ON QUANTUM THEORY

The Quantum Field Theory is a very sophisticated concept. One must stand in awe that human minds could reach into such wonderful complexities and weave a theory out of it which we have at present the most accurate predictor in physics today. What are the mathematical ingredients?

First: there is the Lagrangian L: This is merely the difference between the kinetic and potential energy. When changed to $\mathcal{L}$ it becomes the Lagrangian density.

Second is the Hamilton principle—a general principle which states that any system evolves to minimize L and therefore also $\mathcal{L}$.

Third: Perturbation. This approximates the equations of motion and thus describes how quanta can be created and destroyed. Now we can symbolize:

$\mathcal{L}_0(\psi_e)$ the Lagrangian density of the wavefunction of the electron

$\mathcal{L}_0(A)$ the Lagrangian of the four vector representing the electromagnetic field.

$\mathcal{L}_{int}(\psi_e A)$ the interaction between electrons and photons

Total $\mathcal{L} = \mathcal{L}_0(\psi_e) + \mathcal{L}_0(A) + \mathcal{L}_{int}(\psi_e A)$

Now Feynman comes in with his renormalization—a process which he himself called dippy. With the infinities removed and loaded with Feynman diagrams which are spacetime diagrams showing every possible interaction in a particular situation, you have the elements of QED. One of the many triumphs of QED was to calculate the electron magnetic moment anomaly.

THE ANALOGY WITH MOIRÉ SHEETS.

The two transparencies are really nothing but moiré overlays. They serve here as a comparative analogy of quantum theory and EST. It becomes instantly clear why quantum mechanics are as complex mathematically and difficult to explain. The patterns are here—there and everywhere. They compare well to the electron which in one position of the sheets imitates the electron wave spread across all of space. With a little juggling one can begin to localize them into apparent diffraction lines. A little more juggling and they illustrate a wave packet where the lines are close together.

EST is illustrated by the utter simplicity of the parallel lines on either of the two pages.

CHAPTER IX

UNITLESS IDENTITIES OF NATURE

It has long been the desire of physicists to express natural values by stating them in terms of one another. These natural numbers are often referred to as dimensionless numbers which are independent of our units of measurement. For example we have the ratio of the nucleon and the electron mass as $m_n/m_e = 1836$. The value of the fine structure constant α is a dimensionless number. In the absence of a good dimensionless value for time there was reference made to a "jiffy" which is the time it would take for light to pass the diameter of an electron. The jiffy is quite useless because we don't have a dimension for the electron in modern physics. A good deal of hypothesizing resulted from thinking along these lines—this includes the Large Number Hypothesis which toys with a numerology which purports to show that the universe follows a scheme of groupings which cluster in the so-called unity group and the 10^{40} group of numbers. Some of our greatest minds have been fascinated by the "magic numbers" amongst them being Dirac, Eddington and others. This curious logic has shown that by the manipulation of the numbers, nature has certain preferential groupings but the results actually only led to further dilemmas or else—nowhere!

The entropic space theory, if correct, has yielded a large number of relations hitherto unavailable. By eliminating the constants from many equations, the resulting identities would offer the gist of the equations. The results would not be exchangeable to our measuring system but could be used in relating one dimensionless identity to another.

The primary constants to be removed would be h, c, k_{dil}, π.

Where we have the permeability of space μ_o and the permissivity of space ϵ_o their relation to each other is:

$$\frac{1}{\mu_o \epsilon_o} = c^2 \text{ so that}$$

$$\left| \frac{1}{\mu_o \epsilon_o} \right| = \left| 1 \right| \text{ or } \left| \mu_o \epsilon_o \right| = \left| 1 \right| \text{ and } \mu_o \text{ \& } \epsilon_o \text{ are both 1}$$

$$\text{or } \frac{k_E}{k_M} = c^2 \text{ therefore } \left| \frac{k_E}{k_M} \right| = \left| 1 \right| \text{ and } k_E \text{ \& } k_M \text{ are both 1}$$

CONVERSION OF ENTROPIC VALUES TO UNITLESS IDENTITIES

TABLE 1

ENTROPIC VALUE	UNITLESS IDENTITY
Fundamental charge $q_o = \left[2\ell_o E_o \right]^{1/2}$	$\left| q_o \right| = \left| \sqrt{2} \right|$
Fundamental length $\ell_o = \dfrac{h}{k_{dil}}$	$\left| \ell_o \right| = \left| 1 \right|$
Fundamental Energy $E_o = h(1)$	$\left| E_o \right| = \left| 1 \right|$
Fundamental mass $m_o = \dfrac{h}{c^2}$	$\left| m_o \right| = \left| 1 \right|$
Elementary charge $e = q_o \, \omega_e$	$\left| e \right| = \left| \sqrt{2} \, \omega_e \right|$
Dilated mass $m = \dfrac{h\omega}{c^2}$	$\left| m \right| = \left| \omega \right|$
Dilated length $\ell_{dil} = \dfrac{h\omega}{k_{dil}}$	$\left| \ell_{dil} \right| = \left| \omega \right|$
Dilated charge $q = q_o \, \omega_{dil}$	$\left| q \right| = \left| \sqrt{2} \, \omega \right|$
Fundamental time $t_o = \dfrac{\ell_o}{c}$	$\left| t_o \right| = \left| 1 \right|$

EMPIRICALS

There are some constants in the lexicon of physical constants which are not directly traceable to the constants c, h, k_{dil}, k_F, k_M or π. They have evolved empirically from observations. Among these are N_A, Avogadro's number and k, the Boltzmann constant which is related to N_A and R, the molar gas constant.

G, the supposed gravitational constant which we find not to be a constant and can be expressed in entropic terms as:

$$G = \frac{\ell_p^3 \, c^3}{\hbar} \text{ since } \ell_p = \ell_o \, \omega \text{ we can rewrite:}$$

$$G = \frac{[\ell_o \, \omega]^2 \, c^3 \, 2\pi}{h} \text{ then } \left|G\right| = \left|2 \, \omega^2\right| \text{ after removing } \ell_o, c, \pi, h$$

THE PLANCK EQUATIONS

The equations by Max Planck on length, mass and time, although often quoted have no real application in modern physics. The entropic space theory has given some meaning to these. It would be interesting to see what the results are if the constants are removed and the essence of the equations exposed.

Planck mass:

$$m_p = \left[\frac{hc}{2\pi G}\right]^{1/2} \text{ when reduced to unitless identities:}$$

$$\left|m_p\right| = \left|\left[\frac{1}{2(2)\omega^2}\right]^{1/2}\right| = \left|\frac{1}{2\omega}\right| \text{ (see end of this chapter)}$$

Planck length:

$$\ell_p = \left[\frac{h \, G}{2\pi c^3}\right]^{1/2} \qquad \left|\ell_p\right| = \left|\left[\frac{2\omega^2}{2}\right]^{1/2}\right| = \left|\omega\right|$$

Planck time:

$$t_p = \frac{\ell_p}{c} \qquad\qquad \left|t_p\right| = \left|\omega\right|$$

88

TABLE 2
Well known atomic constants

PRESENT EQUATION	UNITLESS IDENTIFICATION
Magnetic flux quantum $\Phi = \dfrac{h}{2e}$	$\left\|\Phi\right\| = \left\|\dfrac{1}{2\sqrt{2}\,\omega_e}\right\|$
Josephson frequency-voltage quotient $\dfrac{2e}{h}$	$\left\|2\sqrt{2}\,\omega_e\right\|$
Quantized Hall conductance $\dfrac{e^2}{h}$	$\left\|2\,\omega_e^2\right\|$
Quantized Hall resistance $R_H = \dfrac{h}{e^2}$	$\left\|R_H\right\| = \left\|\dfrac{1}{2\,(\omega_e)^2}\right\|$
Bohr magneton $\mu_B = \dfrac{eh}{4\pi m_e}$	$\left\|\mu_B\right\| = \left\|\dfrac{\sqrt{2}\omega_e}{4\omega_e}\right\| = \left\|\dfrac{\sqrt{2}}{4}\right\|$
Nuclear magneton $\mu_N = \dfrac{eh}{4\pi m_p}$	$\left\|\mu_N\right\| = \left\|\dfrac{\sqrt{2}\omega_e\omega_p}{2}\right\|$
Fine structure constant $\alpha = \dfrac{\mu_o c e^2}{2h}$	$\left\|\alpha\right\| = \left\|\dfrac{2\omega_e^2}{2}\right\| = \left\|\omega_e^2\right\|$
Rydberg constant $R_\infty = \dfrac{m_e c\,\alpha^2}{2h}$	$\left\|R_\infty\right\| = \left\|\dfrac{\omega_e(\omega_e^2)^2}{2}\right\| = \left\|\dfrac{\omega_e^5}{2}\right\|$
Hartree energy $E_h = 2R_\infty hc$	$\left\|E_h\right\| = \left\|\dfrac{2\omega^5}{2}\right\| = \left\|\omega_e^5\right\|$
Quantum of circulation $\dfrac{h}{2m_e}$	$\left\|\dfrac{1}{2\omega_e}\right\|$
Electron mass $m_e = \dfrac{h\omega_e}{c^2}$	$\left\|m_e\right\| = \omega_e$
Electron specific charge $\dfrac{e}{m_e}$	$\left\|\dfrac{\sqrt{2}\omega_e}{\omega_e}\right\| = \left\|\sqrt{2}\right\|$

PRESENT EQUATION	UNITLESS IDENTITIES
Compton wavelength $\lambda_c = \dfrac{h}{m_e c}$	$\left|\lambda_c\right| = \left|\dfrac{1}{\omega_e}\right|$
Bohr radius $a_o = \dfrac{\alpha}{4\pi R_\infty}$	$\left|a_o\right| = \left|\dfrac{2\omega_e^2}{4\,\omega_e^5}\right| = \left|\dfrac{1}{2\,\omega_e^3}\right|$
Classical electron radius $r_e = \alpha^2 a_o = 2k_E \ell_e$	$\left|r_e\right| = \left|2\omega_e\right|$
Thomson cross section $\sigma_e = \dfrac{8\pi}{3}r_e^2$	$\left|\sigma_e\right| = \left|\dfrac{8\,\omega_e^2}{(3)2^2}\right| = \left|\dfrac{2\omega_e^2}{3}\right|$

TESTS

The question arises as to whether modern, classical or entropic ways of describing the same thing will result in the same expression of identities. Given below are several tests: The first example consists of four ways to describe the Bohr radius.

$$a_o = \frac{\alpha}{4\pi R_\infty} \qquad \left|a_o\right| = \left|\frac{2\omega_e^2}{4\omega_e^5}\right| = \left|\left|\frac{1}{2\omega_e^3}\right|\right|$$

$$a_o = \frac{h^2}{e^2 m_e k_E} \qquad \left|a_o\right| = \left|\frac{1}{2\omega_e^2\,\omega_e}\right| = \left|\left|\frac{1}{2\omega_e^3}\right|\right|$$

$$a_o = \frac{2k_E \ell_{dil\,o}}{\alpha^2} \qquad \left|a_o\right| = \left|\frac{2\omega_e}{(2\omega_e^2)^2}\right| = \left|\frac{2\omega_e}{4\,\omega_e^4}\right| = \left|\left|\frac{1}{2\omega_e^3}\right|\right|$$

$$a_o = \frac{h}{2\pi m_e c\alpha} \qquad \left|a_o\right| = \left|\frac{1}{2\omega_e\,\omega_e^2}\right| = \left|\left|\frac{1}{2\omega_e^3}\right|\right|$$

Another example may be the Quantised Hall Resistance

$$R_H = \frac{h}{e^2} = \frac{\mu_o^c}{2\,\alpha} \qquad \text{both expressions work out to be:}$$

$$\left| R_H \right| = \left| \frac{1}{2\omega_e^2} \right|$$

Still another example is the Hartree energy E_h which may be expressed as

$$E_h = \left| \frac{e^2}{4\pi\epsilon_o a_o} \right| = \left| \frac{(\sqrt{2}\omega_e)^2\,2\omega_e^3}{4} \right| = \left| \frac{2\omega_e^2\,2\omega_e^3}{4} \right| = \left| \omega^5 \right| \text{ or as}$$

$$E_h = 2R_\infty hc = \left| \frac{\omega_e^5\,(2)}{2} \right| = \left| \omega^5 \right|$$

An interesting observation about unitless identities leads to this rule: *Any relationship given in measurements of the same kind will hold the same relationship numerically by unitless identities.*

For example:

SI Value	**Unitless identity**

fundamental charge q_o = 1.2966903x10^{-39} C $\qquad \sqrt{2}$

elementary charge e = 1.6021773x10^{-19} C $\qquad \sqrt{2}\omega_e$ = 1.7473879x10^{20}

$$\frac{e}{q_o} = 1.2355898\text{x}10^{20} \qquad\qquad \frac{1.7473879\text{x}10^{20}}{\sqrt{2}} = 1.2355898\text{x}10^{20}$$

Another example:

Hartree energy E_h = 4.3597882x10^{-18} J $\qquad \omega_e^5$ = 2.8798613x10^{100}

Rydberg constant $R_\infty hc$ = 2.1798741x10^{-18} J $\qquad \dfrac{\omega_e^5}{2}$ = 1.4399307x10^{100}

$$\frac{E_h}{R_\infty hc} = 2 \qquad\qquad \frac{2.8798613\text{x}10^{100}}{1.4399307\text{x}10^{100}} = 2$$

PLANCK MASS

The EST theory hinges, in part, on the Planck length. It is assumed that since the Planck length and the Newtonian G are linked together, then that length must signify a characteristic of the scalar boson and gravity itself. Because G has been measured a number of times in this century it would offer a current measurement in an expanding universe.

Max Planck also wrote an equation for mass—the Planck mass m_p. Many heads were cracked trying to explain the meaning of that equation and the value of that mass. As most people know, there has never been a particle to correspond to that oversized mass. The EST theory would be amiss not to finally expose the meaning of this mass.

$$m_p = \left[\frac{\hbar c}{G}\right]^{1/2} = 2.17671\times10^{-8} \text{ kg} \quad \ell_p = \left[\frac{\hbar G}{c^3}\right]^{1/2} = 1.61605\times10^{-35}\text{m} \text{ (eq.1a \& 1b)}$$

In this real world there has never been a particle big enough to match this size. So the mystery remains since 1913 when Max Planck wrote this.

The well known relationship for energy and wavelength is

$$E = \frac{\hbar c}{\lambda} \quad \text{or the equivalent mass m} = \frac{\hbar}{\lambda c}$$

If we placed ℓ_p in place of λ we would find that m_p is the monstrous mass.

This suggests that ℓ_p is a wavelength except that we end up with a non-existing mass. The converse is that ℓ_p is not a wavelength, what then is it?

In the model making it became very clear that ℓ is not a wavelength. But it is a length which we call a gravity length. Since $\ell = \lambda y$ only λ should be used to get the mass.

To make this clearer: In (eq.1) Planck used G.

$$G = \frac{\ell_p^2 c^3}{\hbar}$$

Instead of G we can use the value of ℓ_p^2. Moreover we can use $\lambda^2 = \ell^2/y^2$ to isolate the true wavelength λ from ℓ.

The corrected Planck mass m_p:

$$m_p = \frac{\hbar^2}{\lambda^2 c^2} = 9.3903815\times10^{-41} \text{ kg or } 8.439654\times10^{-24} \text{ J}$$

We find that both, dilated mass and energy are in gist $|\omega|$, but the Planck mass is $|2\omega^{-1}|$. This immediately caught our attention. The Planck mass m_p should be just $|\omega|$ like any other dilated mass.

The conclusion is that the Planck mass is non-existent. Using the ℓ_p in place of

λ yields a wrong mass. This is an example of how the gist of an equation can help in finding an incongruity.

What should the mass be, using ℓ_p in accordance with the EST theory?

$$\omega_p = \frac{\ell_P}{\ell_o} = 1.2737044 \times 10^{10}$$

$$E_p = E_o\, \omega_p = 8.4396614 \times 10^{-24}\ \text{J} \qquad \text{or} \qquad m_p = 9.3903893 \times 10^{-41}\ \text{kg}$$

Roughly a ten billionth of the mass of an electron.

E_P The planck energy happens to be exactly $E_{now} \times (2\pi)^{1/2}$ This is the same correction we've given to $\ell_p = \ell_{now}\, (2\pi)^{1/2}$

The wavelength $\lambda = hc/E = 2.3537 \times 10^{-2}$ m

The gravitational length is $\ell_p = 1.61605 \times 10^{-35}$ m.

The amplitude $y = \ell_p E/hc = 6.865983 \times 10^{-34}$

The existence of the amplitude y is directly attributable to model making. Without y we get ridiculous answers for m_p. The Planck length and time remain undisturbed.

CHAPTER X

FORCES

The strong, weak, gravitational and electric forces can be expressed in terms of rest energy E and the following proportional constants:

c, h, k_{dil}, and K_E

Each of the above constants are expressions of the entropic spacetime theory.

In the standard model physics one often hears speculations that these forces were all equal once—when the universe was very young and in the course of its development the forces have diverged.

THE STRONG FORCE:

The entropic spacetime theory points to these forces as all coming from the same source, as the topological models indicate, namely space itself. The apparent diversity of the forces is due to the topological manifestations of the fundamental action h acting ω times; in other words it is due to the energy contained in the "winding" frequency. This is the strong force, the force which is also present in the photon and in all mass particles, but in different quantities. Topologically, the strong force is *twist*. The Strong Force is present in all vector bosons (photons), through phase changes ϕ_1, ϕ_2, ϕ_3 and finally in quarks, neutrons and protons.

In a vector boson of sufficient strength required to make an electron, there is already evidence of the strong force. The electron is created by a collision of the vector boson which brings forth an untwisted loop. The strong force, is substantially diminished by the collision.

The rubber band analogy has the band flat in the scalar state—this does not prevent it from being stretched. The amount of stretch is measured in the amount of energy contained in the band, or the length ℓ of the band. Since the length is proportional to the energy content we derive the constant k_{dil}, the elastic constant based simply on Hooke's law about stress and strain.

The elastic constant is the equivelent of the potential strong force which is present in the scalar bosons of EST.

$$\text{Force} = \frac{\text{mass x distance}}{\text{time}^2}$$

mass, as energy $= E/c^2$
distance is ℓ
time is ℓ/c

$$\text{Force} = \frac{E}{c^2} \frac{\ell}{\ell^2} \frac{c^2}{} = \frac{E}{\ell} = k_{dil}$$

Rest energy $E = F\ell = k_{dil} \, \ell$

k_{dil} is normally given in units of Jm^{-1}. Since however $1J = 1Nm$ one can also say that k_{dil} is a force in Newtons. The interesting point here is that the strong force does not appear ontologically. It is built into the EST as a constant and develops into particles as demonstrated in the model making.

From the model making, which begins with the conversion of the 3D of scalar space to 4D of vector bosons we see that the energy contained in a particle is based on the number of turns (ω) imparted to the scalar space thus transforming it into vector bosons. ($E = h\omega$)

It is not a surprise that the range of the strong force is very limited—it is entirely internal and is due to the energy contained in the "winding" or the frequency of the twist. The most logical equivalent of "range" in this theory is the amplitude y, which is a unitless dimension, but is related to the length ($\ell = \lambda y$). The values for y are given in the TABLES OF EST VALUES, at the end of this book.

This theory asks the quantum mechanics theorists to give up the idea that space is a turbulent continuum in a state of great violence. Instead we should consider it a placid field of energy and charge which stretches and expands as entropic energy is received in finite kernels.

John Archibald Wheeler writes that space is at a high energy level fluctuation of $\sim 10^{94}$ gcm^{-3}. On page 17 of our *COSMOLOGY* we show that the energy in space is $E_o(V_o^{-1}) = 3.6087 \times 10^{84}$ kgm^{-3}, which is just about the same. (Except we are not fluctuating; we are in tension. Tension is potential frequency.)

THE WEAK FORCE

Associated as the weak force is β decay. We'll consider the decay of the neutron. The range of the weak force is limited due to the fact that it is the rearrangement of the phase(3) particles (as well as the quarks which are composed of these) which change the internal arrangement of the decaying neutron to a lesser energy value. (See *Model making*) A change in arrangement is also a change of ℓ and consequently the energy content. In essence, the weak force is due to a topological adjustment of the strong force. (See derivation below)

THE GRAVITATIONAL FORCE

The range of the gravitational force is present throughout the universe, being the cumulative "stretch" of scalar bosons.

The value of Newton's G can be found in the relationship below:

$$G = \frac{\ell^2_{now} c^3}{h} = 6.67259 \times 10^{-11} \text{ m}^3 \text{ kg}^{-1} \text{ s}^{-1}$$

the value of $\ell_{now} = 4.0508359 \times 10^{-35}$ m the same as the Planck "length" multiplied by $(2\pi)^{1/2}$

According to this theory, G is a variable and should increase with the value of ℓ in the EST. This increase of E and ℓ in space should not affect the rest energy of particles. This is because it is the breakdown of particles which increases the energy content of space. The manifestation of gravity is the difference between the ℓ and E of space vs the ℓ and E of the particles.

THE ELECTRICAL FORCE

What about the electrical force? How do we connect this with underlying motivation of all forces, the h and ω? Is it still related to the topological order?

In the topological creation of the electron we have shown that the reason it is electrically discernible is because of the exposure of the charged side to the exterior, while the opposing side is hidden in the interior. This stresses the entropic spacetime and creates a field. Note that the convenient version of the electrical constant K_E contains μ_o, the "permeability" of the entropic spacetime μ_o ($K_E = \mu_o c^2/4\pi$) or is also written as $K_E = 1/4\pi\epsilon_o$ as the ϵ_o is called the "permittivity". Both μ_o and ϵ_o are related to c^2, since $\mu_o \epsilon_o c^2 = 1$. The reason we prefer to use K_E, the electrical constant is because it sits in the electrical equations like Newtons G sits in the gravity equation. (As an aside, you might be amused to see that $K_E/K_M = m/E = c^2$ where k_M is the magnetic constant.)

The model of a photon itself is clearly a model of an electromagnetic wave. The $h\omega$ content of a photon is the measure of a photon. Moreover there is a direct connection between the electrical constant K_E and the elastic constant k_{dil} of EST space:

$$k_{dil} = \frac{F_E}{2 k_E \ell_e} \quad \text{where } F_E \text{ is the internal electrical force of the electron}$$

The derivation of the Electrical Force:

$$F_E = \frac{K_E e^2}{\ell_e} = \frac{K_E 2E\ell_e}{\ell_e} = K_E 2E \quad \text{substituting for } F_E \text{ yields } \frac{E}{\ell} = k_{dil}$$

OTHER DERIVATIONS

Strong Force:

First we shall look at the origin of the Strong Force. This must begin with the force inherent in the entropic spacetime itself—the force which holds space together. From $F = ma$, $m = E/c^2$ and acceleration $a = \ell/t^2$. Since $t = \ell/c$, $t^2 = \ell^2/c^2$. Putting it all together.

$$F_{EST} = \frac{E \ell c^2}{c^2 \ell^2} = \frac{E}{\ell} \quad \text{which is also the constant } k_{dil}$$

Now let's see what happens when our model is twisted; the space is converted to create a vector boson. This photon has a wavelength λ and an amplitude y.

The length $\ell = \lambda y$, where $\lambda = hc/E$ and $y = q^2/2hc$

$$F_s = \frac{E}{\ell} = \frac{E^2}{hc} \frac{2hc}{q^2} = \frac{2E^2}{q^2}$$

The scalar space really has only a field of energy and charge which are real. The length ℓ, the frequency ω and the wavelength λ are merely tools in the concept of two dimensional length separated by time. There is no surface either— only the arrangement of positive and negative charges held apart by E, energy. To create λ, the action h has to act ω number of times in a 2π way—essentially a number of twists.

It is possible to stretch the fabric of space without twisting it. The absorbing of energy stretches the space quanta. The stretch caused by twisting creates ℓ length, λ wavelength and ω frequency. Without twisting they are only potential values. Once twisting occurs the λ and ω are real.

In this sense, the λ length is applicable in the determination of the Strong Force.

$$F_s = \frac{E}{\lambda} = \frac{E^2}{hc} \quad \text{where E/c is momentum and E/} h = \omega, \text{ frequency.}$$

Below are the actual forces within the electron and in the proton.

$$F_e = \frac{8.1871113 \times 10^{-14}}{2.4263106 \times 10^{-12}} = 3.3743047 \times 10^{-2} \text{ N (electron)}$$

$$F_{Pr} = \frac{1.5032787 \times 10^{-10}}{1.33141002 \times 10^{-15}} = 1.1376323 \times 10^5 \text{ N (proton)}$$

$$\left[\frac{F_{Pr}}{F_e} \right]^{1/2} = \frac{m_{Pr}}{m_e} = \frac{\ell_P}{\ell_e} = 1836.152(8) \text{ the well known ratio of proton/electron}$$

The amplitude y_e (for the electron)

$$y_e = \frac{q^2}{2hc} = \frac{(1.6021773 \times 10^{-19})^2}{(2)1.9864475 \times 10^{-25}} = 6.461213 \times 10^{-14} \text{ (unitless)}$$

The range of the electron is currently assigned the value of $\lambda_c/2\pi$, where λ_c is called the Compton wavelength of the electron. λ_c matches exactly with our λ electron wavelength. $\lambda_c/2\pi$ is the radius of our model. y_e is the width of the model band. (unitless)

The Strong Force of a particle increases as the square of the mass m or the square of the gravitational length ℓ of the particle.

A question arises at this point: How is it that the latent Strong Force which is inherent in EST (5.222×10^{11} N = k_{dil}) disappears when we see an actual Strong Force, in, say, the proton of 1.1137×10^5 N?

The answer is in the cosmology. Recall that the creation of matter (or antimatter) occurs only in the contracting epoch. Energy is released from the EST and twists itself into all the examples described in the model-making which is in a completely different (kinetic) environment. The potential λ and ω exist in the EST only to become real wavelengths and frequencies upon release from the EST.

An analogy might be taking a harp which is fully tuned, where all the strings are in high tension and releasing them. One should not be surprised to find twisted up strings of far lower internal forces existing in that condition.

Derivation of the Weak Force:

The weak force is based on the difference between the mother and daughter rest energy of the particles. Let's take, for example the neutron decay.

$$F_w = \frac{E_m}{\lambda_m} - \frac{E_d}{\lambda_d} = 3.1382\times10^2 \text{ N} \qquad \text{the } _m \text{ and } _d \text{ subscripts denote}$$
$$\text{mother and daughter}$$

This can also be written as:

$$F_w = \frac{E_m{}^2 E_d{}^2}{hc} = 3.1382\times10^2 \text{ N}$$

Derivation of Gravity:

For scalar EST, we find that $F_G = E^2/hc$ as the internal gravitational force

This is derived

from $F_G = \dfrac{GM^2}{R^2}$, the Newtonian version of two equal masses

Substitute E/c^2 for M. The value of G from Planck's equation:

$$G = \frac{\ell^2 c^3}{h}, \qquad \ell \text{ and R are both equal lengths}$$

$$F_G = \frac{\ell^2 c^3 E^2}{h\ \ell^2 c^4} = \frac{E^2}{hc} \qquad \text{this is the gravitational force between two}$$
$$\text{adjoining scalar bosons.}$$

The Gravitational Attraction between any two bodies in space

$$F_G = \frac{E_1 E_2}{hc} \frac{\ell_{now}^2}{\ell^2}$$

where ℓ_{now} is the present spacing length of space quanta and ℓ is the distance between the bodies. E is the rest energy of the bodies. (Kinetic motion of the bodies is ignored.) From the above we have a cause for the mechanism of gravity. ℓ_{now} is a tiny length ($4.0508358 \times 10^{-35}$ meters) and is always smaller than any length between bodies. This explains why gravity is always a single directional force and why there is no negative gravity.

The *"fifth"* force:

It is mentioned from time to time that *dark matter* might be a fifth force. In the *Rationale For Dark Matter* we suggest that EST space itself is a good candidate for dark matter because it is dense in energy and ubiquitous, according to this theory. Obviously it obviates the need for a fifth force.

SUMMARY

The entropic spacetime theory has simplified the comparison of the four forces. First we divide the universe into two columns. One is labeled EST POTENTIAL and the other ACTUAL, KINETIC SYSTEM. A single expression E^2/hc fits the Strong Force F_S, the weak force F_w, the gravitation force F_G under the kinetic column. F_G also fits into the EST column. The potential strong force in the EST column is also marked as k_{dil}, the elastic constant, since they equal each other.

The weak force has no application in the EST column and is marked so.

The Electrical force, at first glance resembles the gravitational force the most, but actually differs more. The gist of the matter is this: For gravity we have a variable "constant" G which we translate into ℓ_{now}. The electrical constant is related to c^2, a real constant, which does not permit translation the same way. We're stuck with K_E or its comrade F_E until someone can show how it is a direct derivative of EST space.

The electrical force can most simply be expressed as $F_E = k_E 2E$; in practice we want charge q to show in the relationship.

$$F_E = k_E q^2/\ell$$

F_E falls equally under both columns. The exception is that in the kinetic system column we differentiate between q which is any EST charge vs. *e* which is the charge of the electron. This helps to distinguish the many dipoles hidden in the proton from the single positive charge.

k_E can be written in terms of k_{dil}, E and ℓ

$$k_E = \frac{F_E}{2k_{dil}E} = \frac{F_E \ell}{2E^2}$$

SUMMARY

EST POTENTIAL		ACTUAL (KINETIC SYSTEM)
STRONG FORCE	$F_S = k_{dil} = \dfrac{E}{\ell}$	$F_S = \dfrac{E^2}{hc}$
Weak Force	N/A	$F_w = \dfrac{E_m^{\,2} - E_d^{\,2}}{hc}$
Gravitational Force	$F_G = \dfrac{E^2}{hc}$	$F_G = \dfrac{E_1 E_2}{hc} \dfrac{\ell_{now}^2}{\ell^2}$
Electrical Force	$F_E = \dfrac{k_E q^2}{\ell} = k_E\, 2E$	$F_E = \dfrac{k_E q e}{\ell}$

CHAPTER XI

LAWS OF EST AND A FEW PROOFS

The *TABLES OF EST VALUES* contain the heart of this theory. Relations between these values are laws of this theory. For example the first law states:

THE TEMPERATURE OF A PARTICLE, TIMES THE PARTICLE'S DENSITY, TIMES THE SQUARE OF THE PARTICLE'S FREQUENCY IS CONSTANT.

$$\frac{T\omega^2}{\ell^3} = 2.3496967\text{x}10^{124}$$

This can be easily checked from the Tables. What is unique about this relationship is that in standard model physics there is no value for ℓ either for particles like protons or electrons and certainly not for space itself. The value for T, temperature of electrons and protons etc. can be readily obtained by dividing the rest energy by the Boltzmann constant. The frequency ω associated with these particles is directly found from the wavelengths and such often used nomenclature as the Compton wavelength for electrons or protons.

SECOND LAW

FOR ANY TEMPERATURE THERE IS A FREQUENCY.

$$k_B/h = \omega/T = 2.0836738\text{x}10^{10} \text{ HzK}^{-1}$$

For example the answer to the question "what is the frequency if the temperature is 273 K?" is

$$\omega = (273)(2.0836738\text{x}10^{10}) = 5.6884295\text{x}10^{12} \text{ Hz}$$

On the Tables we don't find a frequency of this size; it is somewhere between the frequency assigned to EST space and the electron. It therefore must be the average frequency which includes some other particles or atoms, or the kinetic condition of such particles.

The kinetic energy is $E_k = 3k_BT/2$. The rest energy $E_r = k_BT$. One might ask from where the 3/2 comes. The E_k equation is applicable to all gases at the same temperature. Relations which apply to kinetics do not stem from EST theory which is preoccupied with rest conditions. For example the classic equation for the velocity of a gas is

$$v = \left[\frac{3k_BTc^2}{E_r} \right]^{1/2}$$

Now this theory states that EST has no velocity (but moves radiation at velocity c) and has a temperature of 1.532247° K. What does the velocity equation do if we place E_{now} in place of E_r? The answer is $(3c^2)^{1/2}$.

The origin of 3 is the result of the complicated particle bouncing in all directions. This is simplified by imagining a gas in which all particles have the same velocity v and are bouncing in 3 directions x,y,z, with 1/6 going in the positive x direction. Since this is not the case in EST then the equation, without the 3, is the proof that energy is transmitted from space quantum to space quantum at c.

We can conclude that for EST, $E_r = k_BT$ and there is no fluctuation, or pressure exerted by EST.

For all gases: $E_r+E_k = E_T$ (total energy) and

$$E_T = \frac{3k_BT}{2} + E_r$$

IS THERE AN INTERACTION BETWEEN MATTER AND EST?

Aside from the virtual particles which quantum theory claims to emanate from the vacuum there seems to be no observable interaction, at least so say most of our physicists today.

We attribute the mechanism of gravity as being the interaction of lesser dense matter with greater density of EST as another example of such interaction but with no more proof than Higgs particles have or virtual particles have.

When we attributed temperatures to different conditions of an expanding space, it was natural to continue assigning temperatures to electrons, nucleons and any particle which has rest energy. These temperatures are not only associated with the Boltzmann constant K_B but directly with the frequency ω. (See Second Law, above.) Since we know the rest energies of all particles we see very high temperatures at the atomic level. (See Tables.) In the stable particles we do not expect these to cool down according to the second thermodynamic law. We know that hot items vibrate (Feynman called it jiggling) but they don't cool off at the rest energy level. If they are excited and contain kinetic energy, then they will indeed cool off—but not below the rest energy temperature.

John Archibald Wheeler had something to say very much to the point. In his Lecture III ('The boundary of a boundary'):

..."Run alongside one of the molecules and ask it what it thinks of the second law of thermodynamics. It will laugh at us. It never heard of the second law. It does what it wants. All the same a collection of billions of such molecules obeys the second law..."

What happens when we apply cryogenic techniques and bring matter to exceptionally low temperatures? Does that mean we drain the atoms of their rest energy? Of course not! we drain it of its kinetic energy. Helium remains helium no matter how cold the environment is made. Strangely, however, it begins to take on bosonic characteristics. This has been proved by experimental work.

Can we prove that there is a relation between the EST and the kinetic energy of, say, a gas in a cubic meter? We will have proved that the Tables are correct and that the actual temperature of a volume is related to EST space and the gas particle.

As one might expect there are some complications which raise their ugly heads before a real proof can be demonstrated.

For example, it is unknown in this theory, whether the properties of EST are uniform throughout the universe or not. In remote space, such as the 3.5 kpc from the center of a galaxy we want to believe that the properties of space are uniform, because of the uniformity of the velocity of bodies rotating at that distance. Is it not our argument that EST is "dark matter"?

In locations near the center of the luminous portion of a galaxy there might be a tremendous flow of energy—much of which may be entropy. This would enlarge the space quanta, increase the E value and the ℓ value, except that this energy is also distributed into the EST at velocity c. This might bring the input-output relation into the relativistic range. None of this is known at this time but might be clarified later.

First, we can write an equation for the Boltzmann constant k_B which is considered at this time an empirical equation not related to the main constants.

$$k_B = \frac{R}{N_A}$$ where R is the gas constant and N_A is Avogadro's number

We can write this:

$$k_B = \frac{2E_o\omega_{now}}{3} = \frac{2E_{now}}{3} = 1.4 \times 10^{-23}$$ where $E_o = h$

The approximation might be due to the 2/3 which is an approximation, as described above and/or other factors which will be touched upon later.

Now we will show that the kinetic energy of a gas $E_k = E_{now}T_k$.

This means that the kinetic temperature of a gas multiplied by the energy of the EST quantum equals the kinetic energy of that gas. The kinetic energy of any gas is conventionally calculated as: $E_k = 3/2(k_BT_k)$. For example:

The kinetic energy of any gas @ 273°K (atm. pressure) is $5.6537945 \times 10^{-21}$ J. by conventional means.

E_{now} x 273 = $5.7753396 \times 10^{-21}$J. Notwithstanding the slight difference, we consider this a proof that there is a relation between the EST evaluation of the present energy content of a space quantum and the kinetic energy of a gas.

A NOTE ABOUT E_{now}

E_{now} is derived from Planck's length (corrected by $(2\pi)^{1/2}$ from $\hbar$ in the Planck equation) through the basic EST equation $E = q^2/2\ell$. Planck's length was assumed to reflect the present condition of the universe because it contains the Newtonian G and translates it into a length. This assumption was purely a *"what if"* attempt.

The value of E_{now} is derived from the basic equation which ties up charge with energy and length. The fact that E_{now} can be related directly to the Boltzmann constant is extremely satisfying. So is the ability to tie E_{now} with the kinetic energy of gases.

It should be made clear that E_{now} is a variable. It represents the amount of rest energy contained in a space quantum today. We acknowledge an expanding universe and expect this value to increase. The Tables give the difference between the value of E_{now} and anticipated value of Eq, the energy of a space quantum at the equilibrium, perhaps 20 billion years from now.

The Boltzmann constant is a well tested number, but it is also empirical and has an uncertainty of 8.5 ppm. G has been measured with an uncertainty of 125 ppm and then only to the fifth significant place. It is not surprising therefore to find inexact matches. An adjustment of ±1% at either end would yield a perfect match.

The question of whether the EST is generally uniform or if it is strongly affected by the proximity of high entropy bodies is answered herewith. It would seem that EST is reasonably uniform throughout.

MORE "PROOFS"

Let's first say that there really is no "proof" for anything which is related to nature! Only mathematicians have proofs; that is because they invented mathematics to begin with.

What we are going to prove is that data which has been collected by direct measurement such as specific gravity combined with atomic mass and the

Avogadro number can be reasonably duplicated from data in the TABLES. The system which creates the EST theory is best reflected in the TABLES, so that verification of the TABLES is a support of the theory.

HYDROGEN: Sp. gr. $.09 \text{kgm}^{-3}$ @ 273 K and atm. pressure.
Atomic mass: 1.00797

$$N = \frac{\rho N_A}{M} = \frac{.09 N_A}{1.00797} = 5.37707 \times 10^{25} \text{ atoms m}^{-3}$$

HELIUM: Sp. gr. $.1785 \text{kgm}^{-3}$
Atomic mass 4.0026

$$N = \frac{.1785 N_A}{4.0026} \quad 2.6856 \times 10^{25} \text{ atoms m}^{-3}$$

OXYGEN: Sp. gr. 1.429kgm^{-3}
Atomic mass 15.9994

$$N = \frac{1.429 N_A}{15.9994} = 5.3787225 \times 10^{25} \text{ atoms m}^{-3}$$

From Tables

$$\left[\frac{V_{now}^{-1}}{V_{pr}^{-1}} \right]^{2/3} = 5.0495 \times 10^{25}$$

A good approximation considering that we are using just the theoretical number of protons and space quanta per cubic meter.

Let's try another—the average spacing of the gas molecules.

HYDROGEN:
av. spacing $\quad d = \dfrac{1}{N^{2/3}} = 2.6494 \times 10^{-9} \text{ m}$

HELIUM: $\quad d = 3.3392 \times 10^{-9} \text{ m}$

OXYGEN: $\quad d = -2.6491 \times 10^{-9}$

From TABLES:

$$\left[\frac{\ell_{now}}{\ell_{pr}} \right]^{2/3} = 2.7 \times 10^{-9}$$

The "what if" assumption that the Planck length equation does indeed indicate that the ℓ value fits the condition of present space. The interaction of space with the kinetic system may not be dynamic but it is at least related proportionally which is all we can expect between a potential system and a kinetic system. Since the present space condition is variable, we can also say that the Newtonian "constant", G is variable and that empirical constants like the Boltzmann "constant", the gas constant R and the Avogadgro number are also variable. No wonder we could never tie them into true constants like h and c.

APPENDIX 1

WAVELENGTH RANGES FOR SCALAR BOSONS, VECTOR BOSONS AND FERMIONS.

λ (meters)	SCALAR BOSONS	VECTOR BOSONS	FERMIONS
10^8	. . . . . .EST @ big bang		
10^6	. . . . Long wave, communication		
10^3	. . . . AM radio broadcasting		
10^2	. . . . Short wave radio		
10	. . . . TV, FM radio		
10^{-1}-10^{-3}	. . . . Radiation background 5.26×10^{-3}		
9.39×10^{-3} present λ in EST	. . . . MICROWAVES		
8.06×10^{-3} Equilibrium (EST)	. . MICROWAVES		

ultimate λ for scalar boson

10^{-4}	. . . . Infrared		
10^{-6}	. . . . Visible light range		
10^{-7}	. . . . Ultraviolet		
10^{-10}	. . . . X rays		
10^{-12}	. . . . Gamma ray . . . . Electron 2.42×10^{-12}		
7×10^{-13}	. . . . phase (1) tubular		
6×10^{-13}	. . . . phase (2) . . . . .gluon		
1×10^{-13}	. . . . phase (3)		
1×10^{-14}	. . . . cosmic rays . . . . . .muon 1.17×10^{-14}		
3.98×10^{-15}	. . . . .quark = 26x phase (3)		
10^{-15}	. . . . .proton, 3 quark + 3 gluon		

TABLE OF EST VALUES

	$\ell = \lambda y = q^2/2E$	$E = q^2/2\ell$	
	LENGTH ℓ meters	ENERGY E Joules	MeV
Minimum	ℓ_o $1.2687795 \times 10^{-45}$	E_o $6.6260755 \times 10^{-34}$	4.135669×10^{-21}
Now	ℓ_{now} $4.0508359 \times 10^{-35}$	E_{now} 2.115509×10^{-23}	1.320396×10^{-10}
Equilibrium	ℓ_q 4.717×10^{-35}	E_q 2.436×10^{-23}	1.537×10^{-10}
↑ MAXIMUM VALUES OF SCALAR BOSONS ↑			
electron	ℓ_e $1.5676911 \times 10^{-25}$	E_e $8.1871113 \times 10^{-14}$	.51099906
tubular	$\ell_{\varnothing 1}$ $5.2497416 \times 10^{-25}$	$E_{\varnothing 1}$ 2.741625×10^{-13}	1.7111873
phase 2 (gluon)	$\ell_{\varnothing 2}$ $6.0852641 \times 10^{-25}$	$E_{\varnothing 2}$ $3.1779688 \times 10^{-13}$	1.9835313
phase 3	$\ell_{\varnothing 3}$ $3.6670056 \times 10^{-24}$	$E_{\varnothing 3}$ $1.9150573 \times 10^{-12}$	11.952842
Muon	ℓ_μ $3.2414876 \times 10^{-23}$	E_μ $1.6928347 \times 10^{-11}$	105.65839
QUARK ($26 \times \ell_{\varnothing 3}$)	ℓ_Q $9.5342146 \times 10^{-23}$	E_Q 4.979149×10^{-11}	310.7739
PROTON	ℓ_{pr} $2.8785203 \times 10^{-22}$	E_{pr} $1.5032787 \times 10^{-10}$	938.27231
NEUTRON	ℓ_n $2.8824877 \times 10^{-22}$	E_n $1.5053507 \times 10^{-10}$	939.56563
$k_{dil} = 5.22240(11) \times 10^{11} Jm^{-1}$		$hc = 1.9864475 \times 10^{-25}$	

$$\left[\frac{k_{dil}}{2}\right]^{1/2} = \text{rest energy, electron, in eV}$$

$K_E = 8.9875518 \times 10^9 = 1/4\pi\epsilon_o$

$k_B = 1.380658 \times 10^{-23} = \text{Boltzmann constant}$

The relation of k_E and K_{dil}:

$k_{dil} = \dfrac{F_E}{2 K_E \ell_e}$ where F_E is internal electrical force of electron

TABLE OF EST VALUES (continued)

	$q = (2\ell E)^{1/2}$ CHARGE in Coulombs	$\lambda = hc/E = c/\omega$ Wavelength meters	$y = q^2/2hc = \ell E/hc$ Amplitude unitless
Minimum	$q_o = 1.2966903 \times 10^{-39}$	$\lambda_o = 2.9979246 \times 10^8$	$y_o = 4.2321928 \times 10^{-54}$
Now	$q_{now} = 4.1399472 \times 10^{-29}$	$\lambda_{now} = 9.3899269 \times 10^{-3}$	$y_{now} = 4.3140228 \times 10^{-33}$
Equilibrium	$q_q = 4.82 \times 10^{-29}$	$\lambda_q = 8.065 \times 10^{-3}$	$y_q = 5.848 \times 10^{-33}$
↑ MAXIMUM VALUES OF SCALAR BOSONS ↑			
electron	$q_e = 1.6021773 \times 10^{-19}$	$\lambda_e = 2.4263106 \times 10^{-12}$	$y_e = 6.461213 \times 10^{-14}$
tubular	$q_{\emptyset 1} = 3.7937874 \times 10^{-19}$	$\lambda_{\emptyset 1} = 7.2455101 \times 10^{-13}$	$y_{\emptyset 1} = 7.2455101 \times 10^{-13}$
phase 2 (gluon)	$q_{\emptyset 2} = 4.3975879 \times 10^{-19}$	$\lambda_{\emptyset 2} = 6.2506825 \times 10^{-13}$	$y_{\emptyset 2} = 9.735359 \times 10^{-13}$
phase 3	$q_{\emptyset 3} = 2.6500049 \times 10^{-18}$	$\lambda_{\emptyset 3} = 1.0372784 \times 10^{-13}$	$y_{\emptyset 3} = 3.5352184 \times 10^{-11}$
Muon	$q_\mu = 3.3127942 \times 10^{-17}$	$\lambda_\mu = 1.1734445 \times 10^{-14}$	$y_\mu = 2.7623699 \times 10^{-9}$
QUARK $26x(_{\emptyset 3})$	$q_Q = 9.7439494 \times 10^{-17}$	$\lambda_Q = 3.9895321 \times 10^{-15}$	$y_Q = 2.389808 \times 10^{-8}$
PROTON	$q_{pr} = 2.9418424 \times 10^{-16}$	$\lambda_{pr} = 1.32141 \times 10^{-15}$	$y_{pr} = 2.1783703 \times 10^{-7}$
NEUTRON	$q_n = 2.9458971 \times 10^{-16}$	$\lambda_n = 1.3195912 \times 10^{-15}$	$y_n = 2.1843793 \times 10^{-7}$

$$\text{charge density} = \frac{q}{\ell} = 1.0219293 \times 10^6 \text{ Cm}^{-1}$$

TABLE OF EST VALUES (continued)

	$\omega = c/\lambda$ frequency Hz	$t = \ell/c$ time interval seconds	$T = E/k_B$ Temperature °Kelvin
Minimum	$\omega_o = 1$	$t_o = 4.2321928 \times 10^{-54}$	$T_o = 4.799216 \times 10^{-11}$
Now	$\omega_{now} = 3.1927028 \times 10^{10}$	$t_{now} = 1.3512134 \times 10^{-43}$	$T_{now} = 1.532247$
Equilibrium	$\omega_q = 3.717 \times 10^{10}$	$t_q = 1.573 \times 10^{-43}$	$T_q = 1.785$ ESTIM.
↑ MAXIMUM VALUES OF SCALAR BOSONS ↑			
electron	$\omega_e = 1.2355898 \times 10^{20}$	$t_e = 5.2292546 \times 10^{-34}$	$T_e = 5.9298619 \times 10^9$
Tubular	$\omega_{\varnothing 1} = 4.1376308 \times 10^{20}$	$t_{\varnothing 1} = 1.7511253 \times 10^{-33}$	$T_{\varnothing 1} = 1.985738 \times 10^{10}$
phase 2	$\omega_{\varnothing 2} = 4.7961554 \times 10^{20}$	$t_{\varnothing 2} = 2.0298256 \times 10^{-33}$	$T_{\varnothing 2} = 2.3017784 \times 10^{10}$
phase 3	$\omega_{\varnothing 3} = 2.8901832 \times 10^{21}$	$t_{\varnothing 3} = 9.6017101 \times 10^{-31}$	$T_{\varnothing 3} = 1.3870613 \times 10^{11}$
Muon	$\omega_\mu = 2.5548073 \times 10^{22}$	$t_\mu = 1.0812439 \times 10^{-31}$	$T_\mu = 1.22610722 \times 10^{12}$
QUARK	$\omega_Q = 7.5144767 \times 10^{20}$	$t_Q = 3.1802716 \times 10^{-31}$	$T_Q = 3.6063594 \times 10^{12}$
PROTON	$\omega_{pr} = 2.2687316 \times 10^{23}$	$t_{pr} = 9.6017101 \times 10^{-31}$	$T_{pr} = 1.0888132 \times 10^{13}$
NEUTRON	$\omega_n = 2.2718588 \times 10^{23}$	$t_n = 9.614944 \times 10^{-31}$	$T_n = 1.090314 \times 10^{13}$
MAX. POSSIBLE	$\omega_{max} = 2.3628413 \times 10^{53}$		$T_{max} = 1.1339785 \times 10^{43}$

$k_B/h = 2.0836738 \times 10^{10}$ HzK^{-1} = ω/T

TABLE OF EST VALUES (continued)

	EST QUANTUM VOLUME ℓ^3 m^3	EST QUANTUM V^{-1} $1/\ell^3$ m^{-3}	EST QUANTUM ENERGY DENSITY $E(V^{-1})$ Jm^{-3}
Minimum	$2.042483 \times 10^{-135}$	4.896002×10^{134}	$3.24412799 \times 10^{101}$
Now	$6.6471266 \times 10^{-104}$	1.504409×10^{103}	3.1825908×10^{80}
Equilibrium	1.0496×10^{-103}	9.5283×10^{102}	2.3211×10^{80} ESTIM.

$$\uparrow \text{ MAXIMUM VALUES OF SCALAR BOSONS } \uparrow$$

electron	$3.8528445 \times 10^{-75}$	2.595485×10^{74}	2.1249525×10^{61}

FERMI COUPLING CONSTANT $1/G_F = \dfrac{\pi E_e}{\ell_e^{\,3}} = \dfrac{\pi k_{dil}}{\ell_e^{\,2}} = 6.6757354 \times 10^{61}$

tubular $\phi 1$	$1.4468176 \times 10^{-73}$	6.9117213×10^{72}	1.8949348×10^{60}
phase 2 (gluon)	2.253400×10^{-73}	4.4377385×10^{72}	1.4102994×10^{60}
phase 3	$4.9309968 \times 10^{-71}$	2.0279875×10^{70}	3.8837123×10^{58}
Muon	$3.4059094 \times 10^{-68}$	2.9360734×10^{67}	4.9702869×10^{56}
QUARK	$8.66672202 \times 10^{-67}$	1.153839×10^{66}	5.7451363×10^{55}
PROTON	$2.3851071 \times 10^{-65}$	4.1926838×10^{64}	6.3027722×10^{54}
NEUTRON	$2.3949827 \times 10^{-65}$	4.1753955×10^{64}	6.2854345×10^{54}

$$\frac{T\omega^2}{\ell^3} = 2.349696(7) \times 10^{124}$$

$$\left[\frac{V_{pr}}{V_e}\right]^{1/3} = \left[\frac{\rho_e}{\rho_{pr}}\right]^{1/3} = \frac{\omega_{pr}}{\omega_e} = 1836.1527$$

INDEX